AN
ALCHEMY
OF MIND

The Marvel and Mystery of the Brain

Diane Ackerman

SCRIBNER

New York   London   Toronto   Sydney

SCRIBNER
1230 Avenue of the Americas
New York, NY 10020

For information about special discounts for bulk purchases,
please contact Simon & Schuster Special Sales:
1-800-456-6798 or business@simonandschuster.com

BRAIN ILLUSTRATION BY REBECCA GODIN

DESIGNED BY ERICH HOBBING

Text set in Adobe Caslon

Manufactured in the United States of America

1   3   5   7   9   10   8   6   4   2

Library of Congress Cataloging-in-Publication Data

Ackerman, Diane.
An alchemy of mind : the marvel and mystery of the brain / Diane Ackerman.
p. cm.
Includes bibliographical references and index.
1. Brain—popular works. 2. Neuropsychology—Popular works. I. Title.
QP376. A225 2004
612.8'2—dc22
2004041621

ISBN 0-7432-4672-1

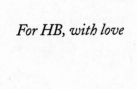

For HB, with love

CONTENTS

PAVILIONS OF DESIRE
(Memory)

THE COLOR OF SAYING
(Language)

THE WILDERNESS WITHIN
(The World We Share)

my mind is
a big hunk of irrevocable nothing which touch and
taste and smell and hearing and sight keep hitting and
chipping with sharp fatal tools
in an agony of sensual chisels i perform squirms of
chrome and execute strides of cobalt
nevertheless i
feel that i cleverly am being altered that i slightly am
becoming something a little different, in fact
myself
Hereupon helpless i utter lilac shrieks and scarlet
bellowings.

                                    e. e. cummings, *Portraits*, VII

MIRACLE
WATERS

(Evolution)

CHAPTER 1

The Enchanted Loom

... an enchanted loom where millions of flashing shuttles
weave a dissolving pattern, always a meaningful pattern
though never an abiding one; a shifting harmony of sub-
patterns.

—Sir Charles Sherrington,
Man on His Nature

Imagine the brain, that shiny mound of being, that mouse-gray parliament of cells, that dream factory, that petit tyrant inside a ball of bone, that huddle of neurons calling all the plays, that little everywhere, that fickle pleasuredrome, that wrinkled wardrobe of selves stuffed into the skull like too many clothes into a gym bag. The neocortex has ridges, valleys, and folds because the brain kept remodeling itself though space was tight. We take for granted the ridiculous-sounding yet undeniable fact that each person carries around atop the body a complete universe in which trillions of sensations, thoughts, and desires stream. They mix privately, silently, while agitating on many levels, some of which we're not aware of, thank heavens. If we needed to remember how to work the bellows of the lungs or the writhing python of digestion, we'd be swamped by formed and forming memories, and there'd be no time left for buying cute socks. My brain likes cute socks. But it also likes kisses. And asparagus. And watching boat-tailed grackles. And biking. And drinking Japanese green tea in a rose garden. There's the nub of it—the brain is personality's whereabouts. It's also a stern warden, and, at times, a self-tormentor. It's where catchy tunes snag,

3

and cravings keep tugging. Shaped a little like a loaf of French country bread, our brain is a crowded chemistry lab, bustling with nonstop neural conversations. It's also an impersonal landscape where minute bolts of lightning prowl and strike. A hall of mirrors, it can contemplate existentialism, the delicate hooves of a goat, and its own birth and death in a matter of seconds. It's blunt as a skunk, and a real gossip hound, but also voluptuous, clever, playful, and forgiving.

The brain's genius is its gift for reflection. What an odd, ruminating, noisy, self-interrupting conversation we conduct with ourselves from birth to death. That monologue often seems like a barrier between us and our neighbors and loved ones, but actually it unites us at a fundamental level, as nothing else can. It takes many forms: our finding similarities among seemingly unrelated things, wadding up worries into tangled balls of obsession difficult to pierce even with the spike of logic, painting elaborate status or romance fantasies in which we star, picturing ourselves elsewhere and elsewhen. Happily storing information outside our bodies, the brain extends itself through time and space by creating extensions to the senses such as telescopes and telephones. How evocation becomes sound in Ravel's nostalgic "Pour une Infante Défunte," a plaintive-sounding dance for a princess from a faraway time, is an art of the brain. So is the vast gallantry of imagining how other people, and even other animals, experience life.

The brain is not completely hardwired, though at times it may seem so. Someone once wisely observed that if one's only tool is a key, then every problem will seem to be a lock. Thus the brain analyzes as a way of life in Western cultures, abhors contradiction, honors formal logic, and abides by many rules. *Reasoning* we call it, as if it were a spice. Cuisine may be a good metaphor for the modishness and malleability of the thinking brain. In some non-Western cultures the brain doesn't reason through logic but by relating things to the environment, in a process that includes contradiction, conflict, and the sudden appearance of random forces and events. The biologist Alexander Luria was struck by this when he interviewed Russian nomads in 1931. "All the bears up North are white," he said. "I have a friend up there who saw a bear. What color was the bear?" A nomad stared at him, puzzled: "How am I supposed to know? Ask your friend!" These are but two styles in the art of the brain. All people are alike enough to be recognizable, even predictable at times, yet everyone has a slightly different flavor of mind. Whole cultures do. Just different enough to keep things interesting, or, depending on your point of view, frightening.

The brain analyzes, the brain loves, the brain detects a whiff of pine and is transported to a childhood summer spent at Girl Scout camp in the Poconos, the brain tingles under the caress of a feather. But the brain is silent, dark, and dumb. It feels nothing. It sees nothing. The art of the brain is to transcend those daunting limitations and canvass the world. The brain can hurl itself across mountains or into outer space. The brain can imagine an apple and experience it as real. Indeed, the brain barely knows the difference between an imagined apple and an observed one. Hence the success of athletes visualizing perfect performances, and authors luring readers into their picturesque empires. In one instant, the brain can rule the world as a self-styled god, and the next succumb to helplessness and despair.

Until now, using the slang we take for granted, I've been saying "the brain" when what I really mean is that fantasia of self-regard we call "the mind." The brain is not the mind, the mind inhabits the brain. Like a ghost in a machine, some say. Mind is the comforting mirage of the physical brain. An experience, not an entity. Another way to think of mind may be as St. Augustine thought of God, as an emanation that's not located in one place, or one form, but exists throughout the universe. An essence, not just a substance. And, of course, the mind isn't located only in the brain. The mind reflects what the body senses and feels, it's influenced by a caravan of hormones and enzymes. Each mind inhabits a private universe of its own devising that changes daily, depending on the vagaries of medication, intense emotions, pollution, genes, or countless other personal-size cataclysms. In Kafka's fiction, a character finds the question "How are you?" impossible to answer. We slur over the sensory details of each day. Otherwise life would be too exhausting to live. The brain knows how to idle when necessary and yet be ready to rev up at the sound of a bear claw scratching over rock, or a math teacher calling out one's name.

Among the bad jokes evolution has played on us are these: 1) We have brains that can conceive of states of perfection they can't achieve, 2) We have brains that compare our insides to other people's outsides, 3) We have brains desperate to stay alive, yet we are finite beings who perish. There are many more, of course.

Sometimes it's hard to imagine the art and beauty of the brain, because it seems too abstract and hidden an empire, a dense jungle of neurons. The idea that a surgeon might reach into it to revise its career seems as dangerous as taking the lid off a time bomb and discovering thousands of wires. Which one controls the timing mechanism? Getting it wrong may be deadly. Still, there are bomb squads and there are brain surgeons. The art of the brain is to liken and learn, never resist a mystery, and question everything, even itself.

CHAPTER 2

This Island Earth

To live is so startling it leaves little time for anything else.
—Emily Dickinson

I was reading this morning about the discovery of a new species of gecko, no larger than a peso, the tiniest reptile on Earth. Found in a sinkhole and a cave in a balding region of the Jaragua National Park, on the remote Caribbean island of Beata, off the southwestern coast of the Dominican Republic, *Sphaerodactylus ariasae* could curl up on the head of a dime and leave room for an aspirin and a deforester's heart. At 16 millimeters (about ¾ inch), it's not only the tiniest lizard, but, according to the biologist Blair Hedges, who discovered it, "the smallest of all 23,000 species of reptiles, birds, and mammals." A female lays but one fragile egg at a time, a minute naïf easily crushed by paws and shoe heels alike in a rain forest more endangered than the Amazon.

Hedges and his colleague Richard Thomas have found only eight of these geckos, and are delighted by their size, but not shocked. They were searching for tiny, overlooked reptiles with limited ranges, because the smallest versions of life tend to inhabit islands. On an island's detached world, over a vast sprawl of time, animals may fill ecological niches snared by others on the mainland. *Sphaerodactylus ariasae* (named in honor of Yvonne Arias, an avid conservationist in the Dominican Republic) is tiny enough, for instance, to compete with spiders elsewhere. The Caribbean is home to many such endangered species, and probably many undiscovered ones that will vanish before they're witnessed or named.

How that saddens me, to think of an animal surviving the rip-roaring saga of life on Earth, minting unique features and gifts, only to vanish without name or record because of human folly. I'm not sure why witnessing a life-form, and celebrating its unique marvel, matters so much to me, but it does. Let's just say it occupies an emotional niche others may fill with prayer. That's a niche shared by many, including the poet Rainer Maria Rilke. In a letter written a year before his death, he speaks of absorbing Earth's phenomena with the full frenzy of human relish and insight as our destiny:

> It is our task to imprint this temporary, perishable earth into ourselves so deeply, so painfully and passionately, that its essence can rise again. . . . We are the bees of the invisible . . . [Our work is] the continual conversion of the beloved visible and tangible world into the invisible vibrations and agitation of our own nature.

Biologists, those "bees of the invisible," had carefully explored the island of Beata, and yet *Sphaerodactylus ariasae* lay hidden for hundreds of years. Could an even smaller reptile exist on Earth? Probably not. There are size limits imposed by gravity and basic biology. But we should always expect the unexpected on remote islands. A century ago, Darwin wrote about the effects of isolation and inbreeding, and how easily island populations diverge from the mainstream and evolve their own genetic dialect. Hence kangaroos live only in Australia (though marsupials abound elsewhere), and hummingbirds only in North and South America (which is why our columbines evolved spurs, unlike their European cousins).

When we become a space-faring species, leaving our home planet to voyage to other worlds, the same fate will become us. Many people won't survive the trips, leaving open niches for stronger, more specialized, or more extreme people to fill. Islands become unique gene pools where uniquely compelled creatures evolve. Multigenerational spaceships, as well as colonies on other planets, if they're not refreshed by outsiders' genes, will function as islands. We may become the bizarre aliens depicted in sci-fi dramas.

Then, although many of *Homo sapiens*' relatives have died out in the past, more will evolve elsewhere, given time's elasticity, and the exuberance of human curiosity. With our restless yen to explore, will our outposts blossom until they're common as pond scum in the cosmic night? I doubt it. But we may become strangers with different sensory talents, develop lizardy skin, evolve into that alien *other* we fear. New habitats will produce new essentials, scarcities, politics, and values. In smaller social groups different dynamics emerge.

That's what happened in our past on this island Earth, and our brains reflect that evolution. Over 500 million years, a span of time too vast to imagine in detail, our brains were molded by environmental stresses and breeding success, while also succumbing to random genetic mutations.

As brains grew, women's pelvises and leg bones widened (hence the characteristic hip swivel). But the skull can only swell so much and still pass through the birth canal. Even after the brain folded in, under, and around itself, it still needed to add important skills. The only solution was to drop some abilities to make room for more important ones. No doubt fascinating gifts were passed up or lost. Based on what other animals evolved, we might have tried sophisticated navigational systems that relied on magnetism or echolocation (like bats or whales). Or a complex sense of smell that made a simple stroll the equivalent of reading a gossip column (like dogs). We might have shared the praying mantis's skill at high-pitched ultrasonics, or the elephant's at low, rumbling infrasonics. Like the duck-billed platypus, we might have been able to detect electrical signals from the muscles of small fish. We might have enjoyed the vibratory sense that's so highly developed in spiders, fish, bees, and other animals. But, of them all, the best survival trick was language, one worth sacrificing large areas of trunk space for, areas that might once have housed feats of empathy that would put extrasensory perception to shame. Indeed, it is possible that people unusually blessed with ESP are lucky ones for whom those areas haven't completely atrophied.

What the brain really needed was space without volume. So it took a radical leap and did something unparalleled in the history of life on Earth. It began storing information and memories outside itself, on stone, papyrus, paper, computer chips, and film. This astonishing feat is so familiar a part of our lives that we don't think much about it. But it was an amazing and rather strange solution to what was essentially a packing problem: just store your essentials elsewhere and avoid cluttering up the cave. Equally amazing was the determination and skill to extend our senses beyond their natural limits, by devising everything from the long eyes of television to the cupped ears of radio telescopes. Forget about being too big for our boots—we became too big for our skulls. Once we imagined gods with supernatural powers, it was only a matter of time before we aped them. On fabricated wings, we learned to fly. With weapons we hurled lightning bolts. Using medicines, we healed. Our ancient ancestors would think us gods.

"Are you out of your mind?!" we sometimes demand. The answer is yes, we are all out of our minds, which we left long ago when our brain needed more room to do its dance. Or rather out of our brain. A born remodeler, it made as many additions as building codes allowed, then designed two kinds of storage bins. Information could be put into things like books that felt good in the hand, and also onto invisible things like airwaves and Internets. "O brave new world," the sea-child Miranda muses in perfect pentameter in *The Tempest*, "That has such people in't." Common sense tells us that if life exists elsewhere in the universe, it will be far more technologically advanced than we. But our evolution has been deliriously quirky, resulting in beings with bizarre traits and personalities, including, for example, the idea of a personality. I wonder how many other planetarians feel the need to share and document their personal existence in such elaborate ways.

The Marine biologist Alister Hardy and the anthropologist Elaine Morgan theorize that at some point in our history, we may have been island exiles much like the tiny gecko. Isolated after a great flood, we might have diverged from other primates into semiaquatic mammals who slept on land but spent most of our days wading or swimming. They point to our loss of body hair; salty tears; layer of subcutaneous fat (something shared by warm-blooded water creatures from ducks to dolphins, but not chimpanzees); flexible spine; streamlined body; seal at the back of the nose to keep water out of the lungs; lowering of the larynx (to allow big mouthfuls of air); heads-up posture; long hair on women (for babies to cling to); blood that's mainly salt water; swimming and diving skills; and voluntary breath control, so important for speech. It's a fascinating idea, and plausible, though I doubt we'll ever know for sure. If only we had home movies of our infancy.

We think of a *human being* as a distinct, definable creature, and its life as complex: "a little gleam of time between two eternities," "a glorious accident," "a tale told by an idiot, full of sound and fury, signifying nothing," "a perpetual instruction in cause and effect," "a flame that is always burning itself out," "a dome of many-colored glass," "a long lesson in humility," "a fiction made up of contradiction," "a fatal complaint and an eminently contagious one," "a play of passion," "a comedy to those who think, a tragedy to those who feel,"* and so on. But we might have been very different animals, with different minds and concerns and mental habits. We are who and what we are only after many trade-offs. If so, how did it happen and what was traded? To begin to answer that we need to be of two minds.

---

*Thomas Carlyle, Stephen Jay Gould, William Shakespeare, R. W. Emerson, G. B. Shaw, P. B. Shelley, J. M. Barrie, William Blake, O. W. Holmes, Sir Walter Raleigh, Horace Walpole.

CHAPTER 3

Why We Ask "Why?"

What is the ultimate truth about ourselves? Various answers suggest themselves. We are a bit of stellar matter gone wrong. We are physical machinery—puppets that strut and talk and laugh and die as the hand of time pulls the strings beneath. But there is one elementary inescapable answer. *We are that which asks the question.*

—Sir Arthur Eddington

Sometimes as the fog of sleep lifts, the mind becomes aware of its traffic. Like commuters on an expressway, messages speed across the corpus callosum, a thick bridge of 200–250 million nerve fibers spanning the brain's two hemispheres. More will follow in a continuous stream of hubbub going in both directions. The brain is a duet of specialists which produces a single experience that's part enterprise, part communion, but all process, all motion.

The right brain is the strong silent one. It can see and act, but not report. Only the left brain talks, and it jabbers all day long, in a self-styled monologue and running commentary on the world, punctuated by conversations with other folk blessed (or cursed) with equally gabby left brains. What's more, the two sides specialize in different facets of mind, with the left excelling at speech and language and the right better at visual-motor skills. Heavy lifting is fine, but don't ask the right brain to solve knotty verbal problems. Which is not to say that the right side doesn't process language—it *does*, but weakly compared with the eloquent feats of the left. Damage the left hemisphere and language becomes a nightmare, especially for men (women generally recover better from left-hemisphere injuries). But people vary greatly and the brain is resilient, so, fortunately, some victims with injured left brains do manage to regain speech. Mind you, that doesn't necessarily mean they can write. As it happens, writing isn't much related to speaking. A relatively recent invention, it's not part of our evolutionary heritage, but more like a sophisticated team sport with changing equipment and rules.

These days, it's fashionable to wear the psychic badge of being "a right-brain person" or "a left-brain person," usually to justify arty behavior, or the lack of it. The left-brain person is supposedly eloquent, analytic, introspective, attuned to details, logical, a problem solver, and good at stories, not to mention alibis. But she tends not to see the whole picture, or do math well, be inventive, or have a strong spatial sense. Jigsaw puzzles are out of the question. Like me, she's probably capable of pulling off a highway to get gas and, afterward, leaving in the wrong direction.

It's the right-brain person who supposedly is intuitive, artistic, musical, looks at the parts and sees the whole, is spatial, recognizes faces, is open to dreamwork and free association, does math, and excels at reading all the nuances of emotion. Though right-handed, I hold a telephone receiver to my left ear (which corresponds to the right brain), maybe because it's then easier to decipher the emotional landscape in a caller's voice.

Of course it's not as rigid as that stereotype. Visual details can often be better on the left, and the slower, more prosodic elements of language better on the right. Most people blend left and right brain use so fluently they're not aware of the divide, or that one side toils silently while the other questions nonstop. Some people use both sides equally, in others one side dominates, and then there are those who are grossly lopsided and make you wonder if they're not actually part android or reptile. But even for them, mind isn't a tug-of-war with the left brain on one side and the right brain on the other, but a collaboration, an open exchange.

More surprising, perhaps, the two hemispheres can differ in their outlook on life, how they feel about themselves, their future, and the

treatment they expect from others. The right side manages negative feelings, the left positive. In studies conducted at the University of Wisconsin, people with very dominant left brains had a
better self-image and tended to describe themselves as optimistic,
happy, confident, enthused by life, not as stressed. They weren't
immune to depression, but once afflicted could recover well. People with very dominant right brains felt worse about themselves,
were more anxious and pessimistic, and they easily succumbed to
depression. Small wonder antidepressants and psychotherapy stir
up the left side of the brain. Our emotional lives walk a tightrope
between the two, finding balance despite tipping a little one way or
the other, or rather *because* of the tipping. Each side seems to play
a critical role in diluting the other. This becomes especially poignant
when someone damages only one hemisphere. Then the opposite
side can run wild, and if it's the right side, the person may suddenly
become uncharacteristically sad, violently anxious, or stewed in negative moods.

That we have brains whose left and right hemispheres specialize isn't unheard of in the animal kingdom, but it's odd. Not just
because animals tend to be symmetrical, and even choose mates
based on how symmetrical (apparently free from genetic mutations) they appear, but because backup systems aid in survival. Our
hemispheres are complementary, more like fraternal twins than like
clones. Why would some abilities be laid down in only one hemisphere?

For decades, Michael S. Gazzaniga, director of Dartmouth's
Center for Cognitive Neuroscience, has been conducting ingenious
experiments with "split brain" people, whose corpus callosum has
been surgically severed to prevent the spread of epileptic seizures.
Such studies offer insight into how the hemispheres divide their
labor. For example, if a split-brain person looks at something only
with the right eye (which corresponds to the left hemisphere of the
brain), he can say what he sees. But show the same picture to his
left eye (right hemisphere) and he'll "see" nothing. The Cheshire
cat effect it's called, after the peek-a-boo cat in Lewis Carroll's
Alice's Adventures in Wonderland. When the right brain is asked to

point to the "invisible" object, the person has no trouble. Many clever experiments with long-suffering volunteers have revealed the left brain to be talkative, the right silent.

In another of Gazzaniga's experiments, a personal favorite of mine, he showed one large picture and four small pictures to each of a volunteer's hemispheres, and then the volunteer was asked to choose the small picture that seemed related to the large one. Neither side knew what the other side was viewing. For a snowstorm scene, the volunteer's right brain had to choose among a shovel, a lawn mower, a rake, and an ax. To go with a picture of a bird's foot, his left brain had to choose among a toaster, a chicken, an apple, and a hammer. Not surprisingly, the right brain correctly chose the shovel for the snowstorm, and the left brain correctly chose the chicken for the bird's foot.

Then things got really interesting. When Gazzaniga asked the subject *why* his right hemisphere had chosen the shovel, his talky left hemisphere answered, but since the split hemispheres couldn't exchange information, it didn't know about the snow scene and had no idea why the shovel was chosen. So it quickly made up a plausible explanation based on the only information it had, that a chicken was somehow involved. The subject reasoned: "The chicken claw goes with the chicken and you need a shovel to clean out the chicken shed." Good answer, if wrong. Gazzaniga calls the left brain the Interpreter, "a device that seeks explanations for events and emotional experiences." When something bad or advantageous happens, it's essential to know why so one can predict and prepare for future events. Mystery causes a mental itch, which the brain tries to soothe with the balm of reasonable talk. The left brain, that is; the right brain prefers to turn a mute eye.

Born fictioneers, all of us, we quest for causes and explanations, and if they don't come readily to hand, we make them up, because a wrong answer is better than no answer. Also, a fast good-enough answer is better than a slow perfect answer. We're devotees of the hunch, estimate, and best guess. I find it hard to watch, say, a David Lynch film like *Mulholland Drive,* which shards into free-associative imagery halfway through, and not try to figure it out.

Critics plague Lynch with "But what does it *mean?*" It's not enough to be startling, beautiful, artful, it has to *mean*, even if much of life simply *is*. Despite knowing that, my left hemisphere, not content to joyously perceive, insists on asking why. A word children use relentlessly and adults continue asking. And so we pass our lives, striving to make sense, even if it produces nonsense, which, of course, *we* never utter, only other people with less-exacting minds. Otherwise, we'd feel at sea, and painfully sure, as the philosopher William Gass says in an essay, that "life, though full of purposes, had none, and though everything in life was a sign, life managed, itself, to be meaningless."

Left brain, right brain, cleft brain. What does this have to do with the dime-size bed of geckos? It hints at our brain's evolution, and why the brain's hemispheres aren't the same. Maybe because of abilities we've *lost*, not gained. In an embarrassment of riches, our brain didn't have space for two of all the faculties available to it. Doling them out to separate hemispheres was its salvation. In further experiments with split-brain patients, Gazzaniga tested the brain's ability to group things visually and found that mice could make simple distinctions humans couldn't. As he says, "That a lowly mouse can perceive perceptual groupings, whereas a human's left hemisphere cannot, suggests that a capacity has been lost." Did the dawn of language, requiring an immense amount of brain space, force out that perceptual knack? If so, what other abilities did we sacrifice to make room for our own private storytellers?

As I move my left hand, the right hemisphere's wand, I notice how two parallel blue veins make a tuning fork at my wrist, and wonder if tops of hands are unique, like fingerprints. Then, picturing fingerprints in my mind's eye, I'm reminded of loopy weather systems, a simile that makes me smile. That reminds me of William Safire's clever book title, *Let a Simile Be Your Umbrella*. My hand rises and my eyes watch it. Anywhere my hand goes steals my attention. We've designed computer programs to be our mechanical shadows: move the arrow or cartoon hand to a spot on the screen, and it's like a toddler pointing to a jar out of reach, grasping it with her gaze.

Just point to something and all eyes follow; pointing and desiring go hand in hand. We are the animals who point and name. Moving my hand to the desktop, I check my calendar, where I'm reminded it's Saturday. A farmers' market down by the lakefront beckons with friends and familiar-looking strangers. Should I invite a girlfriend to meet me there? The last time we spoke, her voice sounded tense. Is she miffed with me for some reason? At the market, I'll find handcrafts from pottery to color-drenched yarn to painted T-shirt dresses (does the artist know that his niece just landed her first nursing job?), and whatever local flowers and produce may be in season, which no doubt will include tiny bouquets of fragrant miniature roses. Restaurant stalls will be offering Thai to macrobiotic cuisine. Fresh mushroom and barley soup, ladled by Robert at Your Daily Soup, would taste great for lunch: slippery mushrooms and chewy barley in broth. An inky scrawl on my calendar tells me it's also time to pay taxes in advance, which I dare not forget. My hand reaches for a cup on a saucer, and I feel the porcelain's sudden coolness as my pointer finger angles through the curved handle and I admire the design of Scotch broom with blue butterflies winding around the cup in a botanical frieze, also noting the residue of mocha foam inside, and hearing the tinkling rattle of the cup being lifted unevenly from the saucer.

Much of what happened in that last free-associative paragraph relied on the right side of my brain, from moving my left hand to picturing the layout of the farmers' market to reexamining the emotion in my friend's voice to the very act of free-associating. I can probably thank the left side of my brain for remembering so many details it can weave together into stories about my life and my environment. Sensory feedback informed both sides of the brain, and I used both to predict the outcome of not paying my taxes. But only we human apes can do what I just did. Because we evolved an extra layer of brain tissue, which we use in unique ways, providing more abilities, we have our own special mental world and the ramparts of thought.

Our current neocortex has two hemispheres, each with four

lobes. Physical sensations stalk the parietal lobe. The occipital lobe gives us sight. We hear using the temporal lobe. The frontal lobe moves our muscles. But none of that commands much gray matter. Most of each lobe is employed in the grand human saga of making associations among events, ideas, personal experiences, strategies, and people. It seems absurd to lump all that tempest together, but we do: *thought*. The word even sounds like a thick knot. Endlessly raveling and unraveling, thought combines colorful yarns to clothe each moment.

CHAPTER 4

The Fibs of Being

Something unknown is doing we don't know what.
—Sir Arthur Eddington

Consciousness is the great poem of matter. But consciousness isn't really a response to the world, it's more of an opinion about it. As miscellaneous as our brain is, with many separate domains, we feel continuous, one mind, one life. How is that possible if the brain's a congress of specialists? The brain is a gifted illusionist. Here's one of its best tricks: I seem to perceive the winter woods today in lavish detail, because whatever I pay attention to looms, furiously present, and saturates my awareness. An ice storm has turned a Japanese maple into a glass figurine. As I caress it with my eyes, the rest of the scene blurs, unless I shift my focus to something else, when *that* leaps into view—a female cardinal with taupe breast feathers and beak orange as candy corn sitting atop a starry fence. I don't feel like I'm looking through a periscope, but glancing outside at nature in the round. A subtle sense of all that's lurking in the rest of the scene lulls me into thinking I'm seeing the yard in a single eye-gulp. That bestows a sense of richness, but my awareness isn't really panoramic. We're more like a pair of binoculars with legs. Many things register in memory's lodging house, whether I checked them in or not. Tomorrow I may recall the geometry of bare limbs against the sky, even though I wasn't attuned to it . . . until now.

The streaming of consciousness is yet another sleight of mind. Afloat somewhere between done and undone, we ride a fluid present from moment to moment. Life feels continuous, immediate, ever unfolding. In truth, we're always late to the party. There's a time lag of half a second between perceiving something and becoming conscious of it. I don't just mean the sort of reflex that makes a hand recoil from a stove before the mind says *too hot!* No, all our conscious acts are afterthoughts. Part of that delay the brain spends primping the order of events, so that the world will feel logical and not jar the senses. It takes time for a perception to reach the brain, for the brain to circulate the news, and the sight or tingle finally to hit conscious awareness. It feels like we sense things and know about them at once, but we don't. Brain time isn't world time. A little offbeat by design, we're willing fools, who would otherwise be late for our date with life. We'd feel like we were constantly trailing half a second behind the world instead of keeping up with it. So there's a butler in the pantry who backdates events. In famous experiments (which I detail in the chapter notes) the neuroscientist Benjamin Libet discovered that the brain processes an action before a person *decides* to act. If we really do have free will, shouldn't the decision to act come first?

For years, Libet's experiments have ignited controversy. Our legal system assumes that adults can choose how to behave, but can they? Or does the brain justify its choices by fooling us into thinking we're free agents? Our days percolate a thick brew of choices obeyed, vetoed, or postponed. Some decisions require arduous thought. That's how it feels, anyway. It doesn't feel like an illusion produced by a brain that, for efficiency or to quash rebellion, lets us believe we're in charge. What's being decided probably determines who's in charge, because a monarch isn't always needed; sometimes a shop boss will do, or just a loud veto. Some days, and in some circumstances, we'll have more or less free will, depending on what's at stake. It was William James who observed that our first act of free will is choosing to believe in free will. We like life to be predictable, so we assume our brain is. Might not be. It might be far more flexible than we imagine, and have a full repertoire of ways to handle problems, depending on their severity or urgency.

All that happens offstage. It's too fussy, too confusing a task to impose on consciousness, which has other chores to do, other fish to fry. We don't like to rouse our slugabed mind unless we need to act or react. Pulling on a sock probably wouldn't require veto power, whereas the impulse to call your boss "a half ounce of insufficiently mobilized prick dust" might. Still, we *feel* like we make all the big and little decisions, as we sail through the narrows of a day, choosing comforts, weighing risks, hatching ideas, generating feelings, adding new plot twists and characters to our life story.

There are many other sleights of mind, involving all the senses, including the carnival of optical illusions favored by brain scientists and magicians alike. Our bodies con us perpetually in a host of intriguing ways. One instance: pain in the heart or other organs is referred elsewhere because there aren't nerves linking organs directly to the brain. As a result, our image of our organs is abstract, and the self one feels so sure of is really the possibility of self, phantom limbs filled with actual limbs, a body image partly imaginary, which often extends to include the family one belongs to, the car in which one drives, the carapace of a house in which one lives.

As I think these words, I hold a pen in my right hand, and look to where ink seeps from beneath my fingertips, skywriting loops and squiggles that linger and *mean*. All the while, I hear these words, which seem to be spoken out loud inside my airy skull. It would take pages to illustrate the hand moves, the ink flow, the pen mechanics, the eyes following that small motion, how the mind communes with itself, why it does so in words. Pages of illustrations? No, it would take the other kind of pages—senatorial assistants or court minions—ages to map the experience to the last detail. Skidding mentally, I think: *In the Middle Ages, the sage pages rampaged in stages . . . on a wise beach.* Instead, I write the word *instead,* am fleetingly reminded of bedstead, *inbedstead,* but brush that aside as another verbal skid, one faintly erotic, and refocus my thoughts on *instead.*

We're not normally aware of such hesitations and detours. The mind feels transparent, an illusion that gives us a sense of control and agency. To think what feels true, little of the backstage action must intrude. Thoughts seem to rise as naturally as bubbles in water. Thank heavens we don't have to supervise each muscle, dredge each memory, sign off on every transfer and exchange. Thank heavens we don't have to coordinate the hundred or so muscles in the throat, face, and torso that work together when we speak. Thank heavens we can leave the Krebs cycle (in which oxygen energizes the cells) to the body's experts, or we'd be fretting over it all day and probably grow so anxious we'd hyperventilate. We thrive on the illusion of spontaneity, of the body working magically, somehow exempt from cause and effect. Truth takes too much time. There's all that messy analyzing, explaining, and verifying to do. For efficiency, the body just cons us and goes about its business, which, occasionally, we glimpse. At the level of flesh and bone, we all coast on innumerable lies, swim in deceits, thrive on amoral cost-benefit decisions. And yet, as lump sums, as selves, we detest lying in others, punish deceit, strive for compassion, and don't like being conned. Go figure.

Thus far, no one has defined consciousness in a completely satisfying way, though many have devoted fascinating books to the subject. "Although it is part of my nature," St. Augustine wrote in the fifth century, "I cannot understand all that I am. This means, then, that the mind is too narrow to contain itself entirely. But where is that part of it which it does not itself contain? Is it somewhere outside itself and not within it? How then can it be part of it, if it is not contained in it?" We continue to ask the same questions today. Philosophers, scientists, psychologists, and poets alike have spent lifetimes trying to describe and define consciousness. Other books offer admirable accounts of past theories about it, so I'll go straight to the current debate.

Here are a few of the main camps. Some people believe that consciousness is an essence given to humans by a deity and includes a supernatural entity like a soul; since it's not physical, science can't understand it. Some believe consciousness is completely physical, a mental state emerging from the neurons, and wonder how and why a biological system gives rise to conscious experience. In this second group, there are those who think a host of separate brain systems (vision, taste, hearing, etc.) build our sense of consciousness brick by brick; those who believe synchronized neurons, acting in unison, reach a critical mass that creates consciousness; those who believe consciousness springs from one specific area (a frontal lobe system?) rather than multiple areas; and those who blend approaches. Some people believe it's a mental state that our sort of brain inevitably creates. Some believe it arises from quantum changes in the structure of the neuron, at the level of subatomic particles, where paradox reigns. Some believe consciousness is physical but that we'll never understand it because a system can't observe itself (how can you be objective about subjectivity? and, anyway, which neural activities produce subjective experience?). Some believe consciousness is physical and knowable but that we're not intelligent enough to figure the brain out, though smarter beings probably could. Separate groups believe consciousness can best be understood through philosophy or psychology or science or literature. Some believe consciousness is physical but we'll only understand it if we can find a way to blend the truths of science, psychology, philosophy, and subjective experiences such as art.

I may have left out a few camps. Some are open-minded, others vociferous, and the field is quickly becoming as abstract, self-enclosed, jargon-ridden, and contentious as a new movement in literary criticism. Everyone seems determined to invent his or her own terms while pointing out the foolishness of everyone else's. Or as the physiologist Bernard Katz put it so well: "Certain scientists would no more use another's terminology than they would use another person's toothbrush." Because defining consciousness is part of the puzzle, it's hard for theorists to agree on the target of their discussion, except perhaps to agree that consciousness is sponsored by the brain. But, generally speaking, there are the consciousness-as-flesh people, the consciousness-as-ghost people, and the consciousness-as-divine people. I suppose there are so many differing views because many are right to some degree.

We may have to accept that some mysteries will remain because we evolved brains specifically designed to hide their workings from us. In any case, we can't completely shelve our subjectivity.

With us life-long, it pleases and defines us in crucial ways, and colors every attempt at objectivity. We need the revelations of neuroscience, but also those of psychology, philosophy, and the arts, which have much to teach us about the subjective experience our brain produces. We're unaware of the urgent board meeting of our psyche, always in session, acquiring information from the world and the body, and running cost-benefit analyses, a labor of multitudes of cells, providing steady feedback to the brain. Instead we feel like solo masters of our fate, captains of our souls, the stuff of homily and poetry.

One problem I find with some of the theories about consciousness is their belief that it floats so far beyond the vigor of matter that there must be a luminous bridge we just haven't discovered yet linking brain processes and phenomenal experience. That seems condescending to matter. We're an arrogant, self-infatuated species, and whatever we argue to the contrary, we do believe we're the pinnacle of life on Earth. Just as every parent has the most beautiful child, we have the most dazzling brain. Consciousness, what could be grander? Surely it's more than a mere brain's squishy parts? But maybe matter isn't as mere as we suppose. We live in a rambunctious, dynamic, to us magical-seeming universe full of recombinable stuff. Consciousness is just one form of mischief matter can create. Quartz is another one. As are Jupiter, cactus, bombardier beetles, college students. Matter has legs, and it dazzles even when inert. If one day we venture beyond our solar system, we may discover some of matter's other nifty tricks.

Also, having studied a little the senses of humans and animals, I'm not convinced plants and other animals are less remarkable than we just because we have self-reflecting minds. What drives a rose to bloom may rally and disturb the rose as much as some of our feelings disturb us. An alligator may lie on a lake bottom with eyes black as sen-sen, her brain pretty much a dial tone, but I'm sure she feels her hormones nagging, I'm sure her matter transfigures her mood. What does that feel like to the alligator? I don't believe we'll ever know, much as we might surmise, based on the primitive brain parts we share and our gift for comparison. What does it feel like to be cold-blooded like a reptile? I've seen alligators regulate their body temperature by lying in the sun on a mud bank with tail and one leg in the water, or warming up by swimming under a scumlike raft of tiny flowers. Our warm-blooded version of that might be adding a light jacket though taking off gloves and hat. We both regulate our body temperature and have other experiences in common. But it's impossible to insert oneself completely into the subjective experience of another person, let alone another species with a different ensemble of senses and instincts.

As much as I treasure our mind's suppleness and high jinks, I don't imagine it's fundamentally much different from what other animals experience in lots of antique ways. Only that it *seems* starkly other because we've adapted to such different habitats, we've evolved a neocortex that likes to ruminate on such things, and our brain provides all we know from birth to death. It seems more complex, and is, because we unfurl elaborate states of being. But that just means our brain is more convoluted, not that there's a threshold past which neurons conjure up something supernatural. Mere matter can be luminous or licentious. Every animal inhabits a different universe. Senses attuned to its unique lifestyle, it perceives only what it needs to survive. Part of the thrill of being human is that we're uniquely ordinary. We share most of our past and biology with Earth's other animals. On the subatomic level, we share our basics with matter throughout the universe, with star hatcheries and space foam. But we're also fundamentally different.

So, although male alligators respond emotionally to music, it's for their own special crocodilian reasons. For example, one evening in 1944 scientists invited a French horn player to serenade an alligator named Oscar. Whenever the musician hit B-flat, the alligator bellowed, as male alligators do as part of their mating display. The same thing happened when a cellist played B-flat. Although there's no report of dancing water surrounding the alligator, that probably happened, too, because part of a male alligator's bellow is subsonic and makes the water leap like frying diamonds.

What happened in Oscar's brain is anyone's guess. But our own reptilian brain works just as mysteriously and mainly swims below our pond of awareness. Given the right note, it makes us bellow without understanding why. Then our higher brain, the story-teller, devises an explanation. Sometimes it's an accurate one, sometimes merely convenient. Just as children like to think they're grown-up and more mature than younger siblings, we like to think we're no longer reptilian. We reserve the word and concept for insults, as in "cold-blooded," "creepy," "beady-eyed," and "thick-skinned." "He was pure reptile," a friend once said, describing a date. "He made my skin crawl." In the oldest swamps of the brain, we have a reptilian logic we've saved for millennia. Why part with func-tions that work so well? When we needed other parts, we added them. But we never lost the reptile brain, inside the brain stem, which responds fast and dirty to any threat, real or imaginary. It's not subtle, polite, or always right, but it doesn't need to be. All it has to do is keep us alive long enough to pass on our genes, and that it does with ferocious and sloppy success. I think a psychologist friend put it well when he said: "Men may be housebroken, but we're still dogs."

The first twitchings of self-awareness may have evolved from working memory, which holds crates of information for *present* use. A feature that evolved late, in the frontal lobes, working memory probably began with simple feelings of familiarity. A woman sees a leaf or berry she's seen before. Her senses report its color, shape, location. Her neural structures allow her, when recognizing stim-uli, to feel a quality of familiarity. She experiences *having seen it before*. Things look the same, there's not a novel food source she must test, no change she must notice and evaluate. If from there we unpack the idea of consciousness a little more, we find a kind of self-awareness that may only exist in humans and in the very highest primates, and probably involves the ability to reflect on oneself by touring the past. Our ancestral lady not only feels self-awareness but can skip back in time to remember and re-create it.

The consciousness debate often hinges on how someone defines consciousness, and of course the erotics of definition making is a special human pleasure. So there are many definitions redefined, redirected, or refuted by subsidiary definitions. Sometimes it seems as if their hidden purpose is to loft humans to a height other animals can't reach. By regarding it as merely *animal,* a relic of our past, we can disown whatever part of our nature offends us. I'll return to definitions in Chapter 32, when I explore whether other animals share our mental cavalcade.

The brain's dynamo runs millions of jobs, by mixing chemicals, oscillations, synchronized rhythms, and who knows what else. It is like looking at a mosaic or a pointillist painting in motion. Study the whole and the parts disappear; study the parts and the whole disappears. Maybe stronger brains will solve that problem in future days. I believe consciousness is brazenly physical, a raucous mirage the brain creates to help us survive. But I also sense the universe is magical, greater than the sum of its parts, which I don't attribute to a governing god, but simply to the surprising, ecstatic, frightening everyday reality we all know. Ultimately, I find consciousness a fascinating predicament for matter to get into.

CHAPTER 5

Light Breaks
Where No Sun Shines

In the unconscious, nothing can be brought to an end,
nothing is past or forgotten . . .
—Sigmund Freud,
The Interpretation of Dreams

Having tidied my frillies drawer, paid a few bills, and read the
current issue of *Cerebrum,* I'm feeling a bit peckish. My
inner campfires need more fuel. Up for several hours, I've been
depleting the body's glucose, a key energy source, and my digestive
system is complaining, which initiates a sense of hunger. My emotions get word. The correct motor responses are chosen. My brain
decides between a snack of two small wafers of 77 percent cocoa bittersweet chocolate, located in a masonry jar on the kitchen counter,
and an organic Gala apple with a handful of walnuts, located in the
refrigerator. There is no risk assessment to do. I feel pretty sure an
irate raccoon hasn't slunk indoors. There is no urgency; I won't faint
if I wait a while longer. In the grand equation of my morning, I
don't risk anything by ambling down the hall to stoke up my furnace
a little more, and there's taste-pleasure to be gained.

When I get to the kitchen, I choose all three—apple, walnuts,
chocolate—put them on a small plate, and begin eating the protein
(walnuts) first, which the stomach may set to work on, busying it,
so that the rest of the snack will take a little longer to digest,
keeping my glucose level more even, thus insuring (I hope) that I

won't be hungry again soon. It's part of my plan to eat healthier and shed a few pounds. That's what I tell myself. On a less available level, my brain has its own reasons, which no doubt include wanting to look shapely.

The brain toils seamlessly, above and below the pond scum of awareness, integrating millions of messages, calculations, appraisals, updates—coming from the round body's imagined corners, as John Donne might say, and from its own mirrored hive, its own bees of the invisible. To its named owner, it speaks in streams of consciousness, image, and back talk. There was a time, Julian Jaynes suggests in *The Origin of Consciousness in the Breakdown of the Bicameral Mind*, when we heard voices inside the head, not backchat from a familiar brain, we thought, but otherworldly beings telling us what to do. These days we're so sure of ourselves that we put on headsets to hear voices on purpose, rightly identifying them as other. But Jaynes speculates that in the days before reflection began, our instincts spoke to us with commands on how to survive. We believed them gods, because they seemed wise, stayed invisible, and yet invaded the mind.

Our states of mind flicker like candlelight, faster than we realize but in familiar ways, fed by a steady glow. Or so we'd swear. Count the moods in a day—ten, forty? But who can count the moods in a minute? A large, solid, sayable feeling like passion may color the body, tinged with unconscious dyes of feeling like nostalgia or guilt. Great art captures some of the hubbub. It's as if the brain were a boardinghouse, where busy tenants recognize one another when they pass on the stairway but rarely converse; and some locked rooms hide lodgers supernatural as shadow, unpigmented, faint as rumor.

Not all of the brain works at the same speed. There's lickety-split reflexes, focused analysis, hurried thought, daydreaming, meditation, intuitive thought, and the slow, patient, meandering thought done by the much-maligned unconscious. Insight roams the sea of the unconscious like the Loch Ness monster, a rumor whose wake occasionally becomes visible, but even then it's mystifying and scarcely believed.

Thinking about something is just one sort of brainwork, which solves some kinds of problems. Other sorts of problems require different tactics and less rush. Maybe years. Let coal lie long enough and it can become diamond. Not all coal, and not without a slow steady pressure. The poet Amy Lowell one day got an idea to write a poem about horses, and then she didn't consciously think about it again. "But what I had really done," she explained, "was to drop my subject into the subconscious, much as one drops a letter into the mailbox. Six months later the words of the poem began to come into my head . . ."

Not willing or pursuing something, but letting a little time elapse, gives the brain a chance to dismiss some bad ideas, free itself from habitual ways of reasoning, try new angles, and wait. Wait for something relevant to enter the mix. Wait for new connections to be made. Nothing may hit consciousness for a long spell. Then, as Guy Claxton explains in *Hare Brain, Tortoise Mind*, "some random daily occurrence serves to remind you, even if only subliminally, of the same word or concept," and "that may be sufficient to tip the scales, and you have the kind of sudden, out-of-the-blue experience of insight to which personal accounts of creativity often refer." Civilization advances as more and more of life's essentials are absorbed by the unconscious. The refrigerator light only has to go on when the door is open. Otherwise we waste energy, and energy is life. We lose some energy when we pursue more. If we lose too much we die. "Operations of thought are like cavalry charges in battle," Alfred North Whitehead writes. "They are strictly limited in number, they require fresh horses, and must only be made at decisive moments."

One marker in childhood is a growing awareness of one's edges. In often-cited experiments with a mirror, children and other animals are tested to see if they regard their reflection as other. Where does self stop and world begin? The body's limits are often explored. We learn early on about outside and inside, that we're apparently separate from the air we breathe, the ground we walk upon, the people we love. We learn that only on rare occasions can we enter another person's body: as lover, doctor, nurse, shaman, mortician, fetus. We don't conjugate or blend. We mix. Or so it seems. In truth, at levels too small for us to notice, we're actually mingling with the elements and other life-forms all the time. But the world is subtle and small, riddled with details, and we're large, lumbering creatures whose senses fix on what they need and ignore the rest. We're selective, not comprehensive. We couldn't handle the full truth.

Because we're social beasts, we unconsciously evaluate the personalities of others, their likely responses, our likely responses to their responses (based on our remembered knowledge of ourself), the possible outcome, which may or may not benefit us. Plus our health, wealth, and the mercurial world's events. That's far too much for any brain to be fully aware of in the seconds required to avoid a fistfight or say the words "I do." We'd stand gibbering in our tracks. So most of our decision making stays sub rosa, just as it does for blue-green algae, newts, hummingbirds, and alligators. And sometimes we're savagely aware of choice and consequence, surplus or famine, and try everything from bluff to compromise before risking the torrential losses of warfare.

One job of the unconscious is to act as a workshop for rough-shaping ideas; crafting notions as new parts or tools become available; storing observations until something relevant appears in the landscape—generally soaking, simmering, and incubating ideas. Gradually, while combing through its inventory, it finds bits and pieces that create a pattern. When it slips knowledge of that pattern to the conscious mind, it's a surprise, like a telegram slid under the door. Since we weren't looking for it, and didn't *know* about it, where did it come from? Out of thin air. We experience that unreasoned solution as intuition or insight. It may be wrong. Intuitions sometimes are. It may point us in a useful direction rather than offer a concrete solution. There are sudden intuitions ("snap judgments"), and there are gradual intuitions based on a slow, playful accumulation of details. Einstein credited his success to such a wordless state of deep play, in which images combined by themselves or at his bidding. Because it wasn't exactly conscious, it wasn't communicable, but he could see it in his mind's eye.

I feel the gentle summons of a forming sensation, a few grains of

sand blown onto my forearm, which is registering the first warming rays of sunrise. I noticed that only because I opened the aper-
ture of my awareness a little wider than usual, unleashed my senses
and allowed them to roam. The sun is layering heat onto my skin
with soft brushstrokes. A sharp line of demarcation runs along the
bone, a shadow ridge. It's cooler and darker where the inner arm
faces my body, warmer where the outward-facing skin glitters in
high-contrast light. We know the rules of sensation. Outside and
inside can't touch for too long without mayhem. We could bake,
dehydrate, burn. I knew that before I found words for it, poor as
those few are, and my body knew it when the sun started licking my
skin, which my brain didn't need to attend to at first, then gradually included among its small worries, things to keep an eye on in
case they fatten into harm. Long before that happens, my brain will
tell its host, its *me*, to desert the soon-to-be-gnawing, subtropical
sun and cool off indoors. The brain understands extreme heat and
petal-thin skin, whose built-in cooling system, sunscreen, and
repair shop are sorely limited. Still, I haven't moved. My brain dishes
out discomfort. If that doesn't work, pain will follow, which usually
does the trick. The brain doesn't start with pain, which requires
extra energy and focus and so is reserved for graver threats. The
brain is like a busy parent: it doesn't need to respond to every utterance of the senses, only the important ones. And sometimes it
doesn't recognize what's important until it's been nagged for a while.

Sensory nagging works so well that I've become more and more
distracted from what I was writing about. Only thirty arbitrary
parcels of planetary flux—what humans call minutes—have passed
since I began the last few pages. It didn't take that long to write
them out in longhand with a felt-tipped pen on sheets of white
paper. I let my senses dally a while on their own, unaccompanied by
thought. And, although I didn't mention it, an exquisite speckle-
tailed bird perched nearby and crooned such lovely songs, full of
trilling and clicks, that I felt pleasure and listened to it for whole
minutes. Then it noticed me and jumped off the railing, fell heavily a few feet, and began to flap. I idly wondered why it had waited
that long to fly. That's when cells on my forearm began to yammer

about sand and heat. I didn't intend for those sensations to loom, grab so much awareness, and detour my reverie. But the brain has priorities. Feel first, think later. Now the sun is threatening my flesh with burn. The brain has seen this before. It judges the heat, predicts the outcome, and jabs me with its own version of a sharp stick until I move indoors.

SWEET DREAMS
OF REASON

(The Physical Brain)

CHAPTER 6

The Shape of Thought

A single neuron may be rather dumb, but it is dumb in
many subtle ways.
                                                —Francis Crick

A stand of aspens in Oregon is reputed to be the largest single
organism on Earth, an underground mass from which over a
hundred thousand trees tower. It flows over hillsides and along valleys, looking like a mob of distinct, separate trees, but combining,
in one network bolder than its sum, an organism that masters
time and space. Conquering immobility, a plant's worst nemesis, it
travels. The trees work in unison, despite the small spaces between
them. Some trees telegraph their mood and news. Under attack,
they send chemical messages to their neighbors, warning them of
danger so they can rally a defense. An individual life with a hundred thousand limbs, this vast organism pales beside the billions of
branching neurons in the brain, which are also separate yet part of
a single invisible indivisible life.

Neurons grow like quaking aspens in the forests of the mind,
sprouting from one matrix, a hidden grove. Unlike other cells, they
don't move or divide. They assume branching shapes from pyramid
to star. Best of all, they talk among themselves, eavesdrop, dash off
messages. For that purpose, they have two kinds of limbs, dendrites
and axons: the former to listen, the latter to speak. Dangling from
a neuron's pouchy trunk, dendrites hear what neighboring neurons
signal through their axons. Like elegant ladies air-kissing so as not
to muss their makeup, dendrites and axons don't quite touch. The

contact happens in less than one thousandth of a second, a spell of microtime powerful as fate. Some neurons sprout only a few dendrites, while others send out many, creating a huge, intricate jungle that can talk, ultimately, with 100,000 other neurons. Some neurons broadcast their news, others just gab with each other. Some connections are nonstop, others relayed. Some connections come with the genetic suit, others are etched by experience. If a man tried to count all those connections, devoting only a second to each, the final sum would take about 32 million years. Even if he were frequently reincarnated, but unevolved due to chronically bad karma, we're talking 44,000 lifetimes.

When shocked, refreshed, or just learning something, neurons grow new dendritic branches, increasing their reach and influence even more. It's as if a neuron sends scouts to many outposts to gather news. The scouts gossip among themselves, too. Based on their reports, a neuron passes messages to other neurons, using its axon, a fine branching limb anywhere from a few millimeters to a meter long. As the axon's tip grows, it senses its environment, getting cues on which direction to turn. Somehow all that nosing around, connecting, and networking orients personality, chisels character.

Neurons speak an elite pidgin neither chemical nor electrical but a lively buzz that blends the two, an electrochemical lingo all their own. To speak, a neuron sends an electrical shudder down the length of its axon in a wave created from the ebb and flow of alternating sodium and potassium ions. A neuron's electricity isn't like a lamp cord's, where current rides a cloud of electrons. And it's not like a sizzle of arcing electricity. More like rolling a stick fast, back and forth, between your palms until a spark jumps. Only eighty thousandths of a volt, a pulse races through the thickest fibers at hundreds of miles per hour, and through the thinnest at a creep. Able to fire hundreds of times a second, one brain cell can rally a mob, until whole neural networks convey a word like *teacher*, a feeling like jealousy, an event like a bike ride.

Every New Year's Eve, selected students in the Ithaca College dorms leave their room lights on or off, window shades open, to create huge glowing numerals. At midnight, on cue, they turn their lights on or off to change the numbers. From miles away, people watch and understand. Worlds within worlds, systems within systems, our neural networks combine to organize, integrate, and encode every kiss, hissy fit, prank, and prayer.

How important *are* the brain salts? One spring day, after strolling along a meandering pathway for only forty minutes in Morikami Japanese Gardens in Delray Beach, Florida, I suffered severe heat exhaustion: spine-crumbling weakness, crabbiness, panting, chills. My conscious mind was an empty cutlery drawer. Wood can warp and skull ache; my brain seemed all downstairs. Higher thought—ideas, creating—floated beyond reach. Although I could understand and answer people in basic ways, I seemed to move in a slow, staggering blur. I felt irritable. Speech was too exhausting to initiate. Suffering from dehydration, my brain struggled with its electrical signals and I felt logy. The idea of food sickened me. I dozed between naps and woke craving sleep. My body was in drought and its delicate salt balance out of whack. Two days, a hospital visit, and many liters of water and Gatorade later, I felt normal again. But it brought home to me the brute power and fragile equilibrium of potassium, sodium, magnesium, and other salts essential to the brain's electricity. Without water, they can't dissolve, and it doesn't take much to disturb them— a water loss of only about 2 percent will do.

Brain cells communicate by sort of shaking hands at hundreds of billions of minute contact points called *synapses* (Greek for "clasp together"), slender channels between neurons. At land's end there's a terminal, where a neuron talks to its neighbor by releasing special molecules, more than a hundred different neurotransmitters, that can drift across the tiny gap and bind to receptors on the other side. The molecules are sometimes pictured as keys, the receptors as locks. When all household keys become electronic, I suppose our antique analogy of a lock-and-key fit will change, but for the moment it's still a sexy image: one shape perfectly machined to fit inside another.*

---

*Plumbers and engineers call connecting parts male and female. Years ago, planning a space station hookup, Russian and American scientists gathered to discuss the logistics. I heard from one of them that a lengthy source of disagreement was that neither country wanted to be the female part!

Traditional locks and keys aren't bendable but impervious metal only villains can pick. We've seen the image so many times in films that it's hard to imagine a soft lock unpuzzled by a uniquely designed soft key. We like creating solid objects with angles and edges, which is why we find the surrealist dreamtime of Salvador Dalí's melting pocket watch so arresting. But on the surface of cell membranes, receptors sit like uniquely shaped soft locks. A soft key with the same shape can open the lock, and that key may be a transmitter chemical or a look-alike drug used to fool the receptor. It's still breaking and entering, minus the villains. We long to jangle a large set of chemical keys. Change the image: the receptors are on/off switches that activate the next cell. Change the image: the key is a car key that starts or turns off the engine.

One molecule won't do. It takes a perfectly timed mob, each binding to its receptor. When the door opens, potassium ions rush out and sodium ions rush in, again creating an electrical charge, this time in the receiving neuron, which passes along the information by releasing a chemical to *its* neighbor, which sparks, and so on, until the message zooms among whole environs of neurons. This one process (*synaptic transmission*) underlies everything the brain does, all of our knowledge, motives, whims, and desires. As *The Dana Guide to Brain Health* puts it: "Ultimately all that we are—all our memories, hopes, and feelings—can be boiled down to the banal transfer of a few ions across the membrane wall of brain cells." Hold that thought like a doubloon.

CHAPTER 7

Inner Space

... life is tolerable only by the degree of mystification we
endow it with.

—E. M. Cioran,
A Short History of Decay

Impossible as it sounds, we have more brain cell connections than
there are stars in the universe. The visible universe, I mean, since
96 percent of the measurable universe is invisible, to us at least.
Linger with that thought a moment, picturing the infinities of
space—a carbon-paper night struck through with countless stars.
Then picture the microscopic hubbub in one brain. A typical
brain contains about 100 billion neurons, consumes a quarter of the
body's oxygen, and spends most of the body's calories, though it
only weighs about three pounds. A ten-watt lightbulb uses the same
amount of electrical energy. In a dot of brain no larger than a single grain of sand, 100,000 neurons go about their work at a billion
synapses. In the cerebral cortex alone, 30 billion neurons meet at 60
trillion synapses a billionth of an inch wide. Only a tiny lightning-
bolt-like apostrophe, and a space essential as the gap between
neurons, stands between *impossible* and *I'm possible*.

To cross or not to cross. Sometimes it takes more than one signal to rouse an idle mind, so the brain nags itself, sending the same
signal over and over. Getting a neuron's attention isn't easy—what
if it's a false alarm? Better to doze a while until the nagging can't be
ignored. Something like *persuasion* finally happens. The neuron gets
excited, joins in, spreads the word. A tribe of neurotransmitters

serve as go-betweens. I picture rush hour on the jammed streets of Manhattan, where cars, trucks, limos, taxis, buses, and people converge, hoping not to collide, while death-defying bicycle messengers weave at speed amid the traffic, sailing through the narrows between truck and car, shifting hips and shoulders to balance as they swerve, one-handed, carrying a parcel under an arm. Sometimes they skim metal or bruise a shin, creating chaos for others, but still they continue at speed, hell-bent for their destination—a single doorway on a tall dendritic building. Once a message arrives, anything or nothing can happen. However enthusiastic, an invitation to join a friend at a nudist camp might not excite you. By disposition, you might not even consider it.

Like hidden Caribbean resorts, synapses can favor excitement or inhibition. Francis Crick puts it charmingly in *The Astonishing Hypothesis:* "It is important to realize that what one neuron tells another neuron is simply how much it is excited." Life is commotion. Four fifths of the neurons in the neocortex favor excitement. The neurotransmitter glutamate feeds that excitement; a small molecule called GABA feeds inhibition. Glutamate and GABA are the speed demons, on call for quick responses to colossal amounts of information. Some slower neurotransmitters include serotonin, norepinephrine, and dopamine, which star in treatments for depression. All sorts of molecules—amino acids, peptides, hormones, and even gases (like nitric oxide)—serve as special messengers. What happens really depends more on the mood of the receptor than the willingness of the messenger. The same messenger can be exciting at one doorway and inhibiting at another, causing wildly different outcomes.

I often marvel at how pills no larger than a hummingbird's eye can produce such dramatic results in a big mammal. Because what happens at the synapses is mainly chemical, not electrical, tiny molecules such as antidepressants and sedatives can insinuate their way in and reshape events. My mother, for example, used to take Valium on occasion to calm her nerves, and it always made her sleepy. That's because Valium belongs to a family of drugs that bind with

GABA receptors, the inhibitory ones, and neurons can become so inhibited you feel dozy.

Our sense of self thrives in those minute spaces, convergence zones where puzzles become pictures and concepts. We haunt the synapses, tiny intersections throughout the brain where traffic stalls, collides, or drives across. The space between two neurons, called a *synaptic junction*, provides a narrows where they meet to send and receive news. It is a small liquid space, as is the air between two whispering lovers, yet so much life happens there. Each junction is a bazaar full of commerce, intrigue, and possibility. In the brain, everything depends on almost nothing, a lively space, the vital channel between neurons.

Throughout nature, space can be a powerful essential. Consider fireflies, twinkling like nomadic stars on summer nights. How can a firefly recognize its own mate with so many glows flicking on and off? Mixing two chemicals to create a cold green light, each couple flashes a secret code during courtship, with the male calling first, then waiting for his partner's reply, a come-hither that invites him down to mate. But their password is an absence, not a presence. How long she delays before flashing is what the male decodes. Alas, some larger females (known as femmes fatales) lure their neighbor's mate by mimicking their code, and then eat the male to absorb a tonic that wards off hungry birds and spiders. And females prefer males with longer flashes. But that's another chapter in the annals of life on Earth, which is storied with precious hollows and channels.

The power of a real and present emptiness is what I'm thinking of, the minute space that thrives between the banks of two neurons. Some popular antidepressants exploit the power of that space by filling it with the neurotransmitter serotonin. Known as SSRIs (selective serotonin reuptake inhibitors), Zoloft, Prozac, and their kin work by "inhibiting the reuptake." I picture that happening at the coastline of two countries. In my mind's eye, they're northern European port cities during the Renaissance, with a river flowing between them, across which cargo must sail. A transmitter chemical like serotonin waits in warehouses on the east coast, and at a signal the boats are loaded and they cross the river. At the other side, locals rejoice at the sight of such riches. At that point the serotonin's job is finished. Shippers carry it back to the original shore, where it's once again stored in warehouses, to await another order to cross.

At trillions of synaptic junctions, crossing molecules carry news, even if the news is that there's nothing new to report. The same old rigmarole is worth reporting, too. To do something is to act, but so is doing nothing; acting and not acting are both choices, even if one thinks not acting is not choosing.

A common myth about the brain is that it's as unyielding as a steel vault. But whenever we learn something, the brain mints new connections or enlivens old ones along familiar pathways. One famous example: brain imaging shows that expert violinists develop more motor cortex for the busy left hand than for the right. Neuroscientists like to say that the brain is *plastic,* a curiously inorganic if fashionable word that takes us back to the heyday of plastic in the 1950s, when it entered everyday life as a moldable wonder. The brain bends, learns, effloresces, adapts. When we learn something, we grow new synaptic connections. Neuron trees grow new twigs along their branches, while some of the branches themselves become stronger. Our brain can rewire itself. We do it all the time when we master ice-skating or learn surgery or pick up knitting. Otherwise, how could we have survived the Ice Age, when, among other useful things, we invented the needle? Much of the brain's wiring takes place after birth. The ultimate immigrant, a brain travels light, gathering notions when it arrives. The destination is childhood, a world that begins in the family, where what a child hears, sees, and feels partly designs the growing brain. "The child is father of the man," William Wordsworth writes in his poem "My Heart Leaps Up," and to some extent that's true. We arrive in this world clothed in the loose fabric of a self, which then tailors itself to the world it finds.

As the developing brain blooms and prunes connections, it has to decide which ones to fix permanently in place and which to dissolve. Preserving what's useful and killing the rest, it chooses.

How does it know what's useful? Whatever we use most. Hence the popularity of bad habits. Breaking them feels like splitting welded steel, and in a sense it is. The *Use it or lose it* axiom has a dark side. Behave in a certain way often enough—whether it's using chopsticks, bickering, being afraid of heights, or avoiding intimacy—and the brain gets really good at it. One can master unfortunate skills that are hard to forget. Great for knowing how to protect oneself, balance a bike, or drive a car. Not great when it comes to defense mechanisms still in use long after the threat that created them has vanished. The reason psychotherapy takes time is that the brain has to be retrained at the level of the synapses. One paradox at the heart of all living things is their ability to change while remaining the same. Our minds remain reasonably stable and effective for an entire lifetime, despite all the daily stresses they encounter. And yet they can bend and adapt and revise themselves when necessary.

Like stars in the universe, neurons don't occupy all of the brain, most of which is water. And neurons don't act alone. Though they are little-known players in the drama of the mind, 90 percent of the brain's cells are spidery *glia* (Greek for "glue"). A varied crowd of cells with many jobs, from cook to bodyguard, they're dominated by star-shaped *astrocytes*, which unfurl long arms and reach right into synapses, altering events. Without glia, neurons would be nothing. On their own, neurons can't feed or sheathe themselves, avoid saboteurs, make themselves understood.

For some while, these glial cells were regarded mainly as filler, the gummy sludge holding neurons in place. But now they're taken more seriously, not just as the neurons' servants but possibly their handlers. A tightly packed corps, they nourish the neurons with lactate (manufactured from glucose in the blood). To protect neurons, they can flatten capillaries with the palmlike ends of their tendril arms and thus hold toxins at bay outside the brain. They can clean up spills of glutamate, the vital neurotransmitter that's nonetheless poisonous in excess (it's implicated in stroke and Alzheimer's, for example, and contributes to the headache from MSG in Chinese cuisine).

But glia are also manipulative cells that can converse among themselves, listen to neurons, voice their own concerns, and ultimately influence what neurons say. They may prompt neurons to create more synapses, rouse sleepy neurons and put them to work, and order neurons to strengthen or weaken their best contacts. They may be vital to memory and learning. Glia with many faces and jobs touch neurons, profoundly altering their fate. Coexisting, as they must, both neurons and glia are dependable, dependent, full of talk and back talk, central to the brain's social fabric and perpetual hum.

CHAPTER 8

Attention Please

Steep thyself in a bowl of summertime.
—Virgil, *Minor Poems*

As a tiny jet takes off at the local airport, its wheels spring back like a horse giving a kick over a hurdle, and then it climbs even faster on polished air. While that drama steals my attention, the airport, the wooden bench my bottom sits on, the hunger pangs making radio wheeze in my stomach, the tinkly twitter of a lone goldfinch, and the rest of life recede. Thought evaporates. I become eyes and a mental notepad. I'm engrossed in the jet's climb not because I need that memory to survive, but because novelty is riveting. Before the brain can judge what's important, it must identify the new, strange, different. I know about planes, but this mini-jet, a pocket rocket, I haven't seen before.

The scene might be only a momentary distraction on a dozy morning, a forgettable flutter. Or watching the jet take off might feed me a fact that slinks into memory, like this one: as a plane retracts its wheels, it speeds up and becomes more steerable. Wheels muddy the smooth flow of air. So do birds' legs. Without such clutter, the wind pours like satin above and beneath the wings, designed with a top curve to speed the airflow, creating a vacuum on top that lifts the plane up. As the jet becomes sleeker, it races faster, steadier, and higher.*

*Bird wings work differently: tiny windows in each wing open and close as they flap, letting air through only on the upstroke.

In *Roderick Hudson*, Henry James writes: "True happiness, we are told, consists in getting out of one's self; but the point is not only to get out—you must stay out; and to stay out, you must have some absorbing errand." An absorbing errand as simple as becoming aware of each breath. All forms of meditation are simply ways of paying close attention. Entice the brain to pay attention, and the newsy, noisy self drains away below the thought horizon like a molten sun at dusk. One can lose one's self while listening to a mockingbird's stolen medley, or staring at a stapler's tiny fangs. Not all of the self. The unconscious goes about its chores, runs the blood factory, conducts sub-rosa board meetings of the psyche, and protects its fragile marshes where flocks of feeling, thought, behavior, and belief all roost.

Heed change, life demands, because it's elemental to survival, especially an erratic or peculiar change. *Just attend*, the brain says. *Notice anything new. Something will matter.*

One day a stone-age tribe in Papua New Guinea greeted a charter pilot with bananas for his airplane and a desire to know what sex it was. The plane's wheels were the first wheels they'd ever seen, and the huffing, twirling sky beast had their urgent attention. Novelty ignites the senses. Learn something new (or someone new) and you discover an avalanche of details. But soon the brain switches to a kind of shorthand. Once the brain perceives something, it's primed to recognize it faster the next time, and even faster after that, until it needn't look at it carefully again. Then, as Hegel says, "the known, just because it is known, is the unknown." Knowing people better, you notice them less. What we call boredom is a form of mental abbreviation. Boredom develops as a kind of waking slumber. Unless things change. Or unless we choose to revive some of the sharp sensations we felt earlier but lost when their startling shine began to dull. Piggybacking on a child's discoveries, or enthralled by an artist's informed innocence, we pay fresh attention to what's grown stale, scrape some of the rust and lichen off the brain, and find the world renewed.

Some years ago, I taught a class of writing students whose work was surprisingly jaded and featureless. Where was the texture of life,

I wondered, the *feel* of being alive on this particular planet? Didn't it strike them as astonishing that they shared the planet with goldfinches and heliarc welders and dung beetles and blood brothers and shiitake mushrooms? Where was their fascination with the world pressing indelibly on what they wrote? Most of the students weren't even twenty-five; how could life already have bored them?

One afternoon, I suggested we begin class at the large open window, by enjoying the phenomena visible at that moment, which included lens-shaped clouds signaling high winds aloft; slate shingles on the library tower overlapping like pigeon feathers; magnolia buds burgeoning into fuzzy-coated hulls that looked like fledglings almost ready for flight; half a dozen dog, squirrel, and bird dramas; and many human pantomimes, as small groups of students coalesced and drifted apart. Everyone had to choose one sensory event that seemed eloquent. For a few minutes, we stood quietly and paid attention.

I wondered if I could reacquaint them with a cunning we inherited from our ancestors: we can seize a phenomenon with mental pincers and stop the world in its spin, if only briefly. Look patiently, affectionately, at anything, gather six or eight perceptions, and it will never look the same again. Because Federico García Lorca wrote "A thousand glass tambourines / were wounding the dawn," we know he once sat and watched a crystal sunrise jingling with color as splinters of light reddened the horizon.

We can't enchant the world, which makes its own magic; but we can enchant ourselves by paying deep attention. My life had been changing, I'd been near death several times, and the simple details of being had become precious. But I also relished life's sensory festival, and the depot where nature and human nature meet. Everything that happens to us—from choosing the day's shoes to warfare—shines at that crossroads.

We don't just regard breaths, objects, and nature, of course. We were emotional beasts long before we were thoughtful ones. A palette of primary emotions guided our distant ancestors in most situations, and they still do. It's a productive if sloppy process. The brain attends to a feeling, is distracted by events, ruminates on other things, associates, follows a tangent, returns to the first with additional insight from its travels, perhaps putting it in a wider perspective, perhaps reevaluating it, then moves on, repeating the process in a slowly opening fan of thought. The brain isn't tidy or linear enough to corral all of its ideas about something. Otherwise, I would have settled everything remotely related to memory in the Memory section of this book. The brain notices, feels, learns, moves on, notices more, learns more, feels more, refines its picture based on what it has learned, moves on, endlessly.

What matters is change, the shifting strata of clouds or society. As Emerson says in "Self-Reliance," one of his loveliest essays, power stops in quiet moments. It thrives on the change "from a past to a new state, in the shooting of the gulf, in the darting to an aim." At great cost, we pay attention, scouting for change. Change leads to actions bold as a dash from a burning barn, as calculated as the reappraisal of status. It always rouses the brain from rest or distracts it from other business. Everything was safe a moment ago, as proven by the fact that nothing bad happened, but if something changes, however small, my safety must be reassessed. Life becomes a lost archipelago, islands of safety, most barely conscious, all vanishing behind us as we focus on a newly seen sliver of coral sand within reach.

We share this instinct with other animals, which is why it's wise not to make eye contact with an aggressive dog you're biking past. When it notices you've noticed *it*, that you're paying attention, you become a greater threat, just as the dog seemed a greater threat when it noticed you and started barking. Or you could stop, leap from your bike, which you then use as a barricade, and bark at the dog in a glowering alpha dog tirade of: "No! Go home! Bad dog!" The dog wouldn't feel rebuked or ashamed, of course, but it might defer to your louder threat.

No time is more alive than the intimate now, where truths are eternal. How long is a *now*? Now is everything the mind and senses can cram into about one-tenth of a second. In that tiny lagoon, any news arriving from the outskirts and inskirts of the body feels like a single moment, a right now. Any novelty can distract an animal from whatever it's doing. Chewing and digesting stop. Instinctively it turns toward the culprit and becomes rigidly aware. It loses sight of everything else for about half a second in what's known as an *attention blink*. Even slugs do it. After a garden talk the other evening in which I confessed to liking many things about slugs (for example, their yen to mate at the end of slime gallows), a man shared a curious slug story of his own with me. He'd been doing construction work when a tractor overturned, trapping him under it. While he waited for help, he noticed a small movement nearby and, turning his head, saw a slug standing up like a tiny giraffe, raptly watching him.

This orienting reflex alerts an animal, cramming its senses with new information, while blocking previous plans or activities. Many things can trigger the reflex—surprise, novelty, sudden change, conflict, uncertainty, increased complexity or simplicity. *Be prepared* is its Boy Scout motto. *An emergency may be looming.* An instance of this used to happen regularly in the Falkland Islands, home to penguins and the RAF airbase Mount Pleasant. Crews discovered that whenever they flew over a penguin colony, the resident penguins would all look up, turning their heads to keep the plane in sight. It was irresistible: the pilots soon began flying out to sea, making a tight turn, then flying above the penguins, whose bills pointed up more steeply as the jets flew overhead, until suddenly the penguins would topple in unison.

Novelty excites by nudging us off balance and weakening our stranglehold on habit. An urgent need arises to improvise new skills, learn new rules and customs. This is especially true of mild novelty, when things change only enough to be noticeable. Complete novelty can seem absurd, something to ignore. But partial novelty makes sense up to a point and yet requires a bright response, so it must be taken seriously. Our lidless curiosity, as well as our passion for mystery, exploration, and adventure, springs from this basic reflex. Once an animal becomes curious it grows alert, and that arousal doesn't quit until it explores the sensory puzzle and can assure itself that all is well, nothing much has changed, no fresh action is required. That repeated pattern of arousal, tension, fear, and suspense, followed by a feeling of safety and calm, provides a special kind of pleasure shared by animals the world over. That we enjoy such tidy escapades enough to excite or scare ourselves on purpose hints at what connoisseurs of pleasure and pain we've become. Rapture always begins with being rapt.

A herd of primates has gathered at a watering hole to drink and socialize. One female sitting in a group of several others notices her mate flirting with a receptive female a few yards away. Dividing her attention, she switches back and forth, from her circle to her mate, able to pay attention to only one conversation at a time. His face says he's altogether too interested in the female draping herself around him in provocative ways. Twitching with jealousy, his mate listens hard and glowers and is almost ready to bound over and stake her claim, but for the moment she keeps an eye on the pair, while halfheartedly grooming a neighbor in her own group.

This is called a cocktail party, and it's almost always a scene of divided attentions, especially among couples.

There are many towpaths of attention, along which the mules of worry plod, and they can change with age. Young children seem infinitely distractable, with attention spans short as a drawbridge, because the reticular formation, a brain part needed for paying attention (and also for filtering out lots of unnecessary data), doesn't finish developing until puberty. A lover pays attention to the beloved with shared life and limbs, soulful passion, passionate soul. That usually rallies one's full, doting, if inaccurate attention. Inaccurate is okay among lovers, where it's sometimes best to blur the details a little. As W. H. Auden writes in "I Am Not a Camera,"

> lovers, approaching to kiss,
> instinctively shut their eyes before their faces
> can be reduced to
> anatomical data.

With advancing age, splitting one's attention becomes harder, as does sifting through warring stimuli. Our filters begin to falter and more sensory noise seeps in, which we find distracting and confusing. A prime example of this is noisy restaurant syndrome, when ambient conversations drown out people at one's own table. But other attention skills can take over. I remember the entomologist E. O. Wilson explaining that when a fishing accident blinded him in one eye, he sadly abandoned his plans to study big animals. But, as he began to focus differently, paying loving attention to the close and small, he developed his famous passion for ants. What we pay attention to helps define us. With what does a man choose to spend the irreplaceable hours of his life? For Wilson, it's ants. For another, it might be the entrails of pocket watches.

Worrisome for people who talk on cell phones in heavy traffic, an MRI study of multitasking (the polite word for attention bingeing) reveals that paying attention to two things at once doesn't double the brain's output but lowers it, shortchanging both. Try pulling an extra wagon uphill. When important things clamor for attention, the brain savors the adages *One thing at a time* and *Divide and conquer*, which is probably why brains devised them in the first place. A focus precise as a single coffee bean works best to store memory. Or noting how coffee beans are shaped like tiny vulvas. But I digress. Ignore that. Ignoring something doesn't mean it won't register, since subliminal images wing in silently and roost in shadows.

Somewhere I left a bowl of Reliance seedless grapes, whose taut red skin bursts open with a melon-flavored gush, and I'm hungry for them. But I was mentally composing this paragraph as I removed the grapes from the refrigerator, dangled them into a bowl, washed them under the faucet, and carried them. That's the last I remember of their travels. I'm not sure where I left the bowl *without thinking*. Is it still in the kitchen? Did I set it down on a shelf when, eyeing a book on memory, I paused to read about absent-mindedness?

CHAPTER 9

A Passion for Patterns

Truly, man is but a passing flame, moving unquietly amid
the surrounding rest of night; without which he yet could
not be, and whereof he is in part compounded.
　　　　　　　　　　—George MacDonald, *Phantastes*

The brain is a five-star generalizer. It simplifies and organizes, reducing a deluge of sensory information to a manageable sum. From that small sample, the brain produces an effigy of the world, whose features it monitors. Anything that doesn't fit, or signals trouble, draws a response. As it learns, it compares new phenomena and experiences with old ones. But individuals and events are never identical, only similar in vital ways. The brain doesn't have room to record the everythingness of everything, nor would that be a smart strategy. Exactly remembering a lion only prepares you for the next lion. Instead, the brain files away a sea of clues, alert to the subtlest insinuation of a pattern. The generalizing brain casts a wide net, allowing key features of a wren to predict those of birds in general. Surviving one cliff is enough to predict danger at other rock forms, even if they're not quite the same shape or composition.

The brain is a pattern-mad supposing machine. It maps the known world. Given just a little stimuli it predicts the probable. When information abounds, it recognizes familiar patterns and acts with conviction. If there's not much for the senses to report, the brain imagines the rest. That does mean gambling, sometimes with ruinous consequences. But if we didn't let imagination fill in the blanks, we'd be unable to survive all the novel predicaments and landscapes we encounter. Imagination is a wild card we use in many ways, sometimes just for fun, although it probably evolved to help us anticipate trouble. Limited to our senses, we'd be confined to daytime and familiar environments. As much a confection as our mental maps are, they allow us to predict, rehearse, and make plans.

Many animals scout their surroundings, predict what may happen, and bank memories for future use. Not all. Blue-green algae doesn't need nerves or a brain. Nor do fungi, lichen, trees, flowers, and other plants. Most of life on Earth does fine without a brain. Clever as plants are—and they're diabolically gifted at manipulation, aggression, seduction, deceit, communication, defense, exploitation, and barter—they don't travel much. Despite the Scarecrow's fancy footwork in *The Wizard of Oz*, if you travel any distance, you have to be able to plan and predict, and that requires a brain. In animals that don't travel far—spiders, for instance—the brain can be tiny. In the small brains of birds that navigate over long distances, one finds bits sensitive to the Earth's magnetic fields. In animals like us, who roam great physical, social, and symbolic distances, the brain must be big enough to map the changing world, provide a sense of control, and adapt to those changes. Otherwise we'd leave a scatter of lost selves babbling in the dust beside the road.

Pattern pleases us, rewards a mind seduced and yet exhausted by complexity. We crave pattern and, not surprisingly, find it all around us, in petals, sand dunes, pinecones, contrails. We imagine it when we look at clouds and driftwood, we create and leave it everywhere like tracks. Our buildings, our symphonies, our clothing, our societies—all declare patterns. Even our actions. Habits, rules, rituals, daily routines, taboos, codes of honor, sports, traditions—we have many names for patterns of conduct. They reassure us that life is stable, orderly, predictable. Perhaps because we are symmetrical folk on a planet full of similar beings. Symmetry often reveals that something is alive. For example, the doe standing at the bottom of the yard blends perfectly into the winter woods. Her mottling of white, brown, and black echoes the subtle colors of the landscape. No doubt she browsed there for some time before I detected her. What gave her away was the regular pattern of legs, ears, and eyes. Then the word *deer* leapt through my mind, and I retraced her with my eyes, this time picking out flanks and nose. *Deer!* my mind confirmed, checking the pattern. Patterns charm us, but they also coax and solicit us. We're obsessed with solving puzzles; we will linger for hours before a work of abstract art, waiting in vain for it to reveal its pattern.

In the exquisite rose garden of the Rodin Museum in Paris, where I'd gone to record some comments about Rodin's art, I stood close to his statue, *The Thinker,* and waited for a soundman to adjust what looked like a ferret climbing a pole. For a minute, we all froze so that he could sample the *wildtrack* or *room tone,* the barely audible noises that make up a background sense of "quiet."

The brain has a wildtrack, too. According to Francis Crick, "When nothing much is happening," a neuron sends slow, irregular spikes down its axon at "a 'background' rate, often between 1 and 5 hertz. . . . This continual 'nervy' activity keeps the neuron alert and ready to fire more strongly at a moment's notice." As it becomes excited by incoming signals, it fires faster, at 50–100 hertz or so, and for short spells it can fire a hundred times faster than that.

At last the soundman floated the furry microphone overhead, and I began speaking about Rodin's sensuous sculpture *The Kiss,* just as a church bell struck. The soundman stiffened like an alert stag. The church bell chimed once more. When a full second passed without a third chime, the overhead ferret withdrew and I stopped. One chime is a distracting instance, two may be a coincidence. It would take three chimes for listeners to detect a pattern, something familiar in a chaotic world: church bells. The same is true of birdcall. If a grackle calls hoarsely only once, recording stops. Three calls allow a pattern to build and recede into the background. *More birds, big deal,* the brain says. *What was that about lust in bronze?* At a restaurant in Jamaica, a mistake on the menu offered diners "Steak grilled to your own likeness." What are the odds of a steak with the profile of Eleanor Roosevelt? It could happen. Two such silhouettes? Weird but possible. If a third one appeared, you'd start getting mighty suspicious. We're so in love with patterns that we obsessively create our own, often in threesomes, such as morning, noon, and night; *Macbeth*'s three weird sisters; the three wise men; ready, set, go; a sonata's three-part form; the genie's granting three wishes; small, medium, and large; ABCs; Goldilocks and the three bears; the three little pigs, and so on. Three seems to be our pattern of choice.

From fairy tales, Grimms' or the family's, we learn what repetitions to expect, and how to bear suspense for longer and longer spells, before the endgame with its moral and meaning. There's always a moral or meaning, which in a fairy tale may be that goodness prospers, whereas in a family it may be that narcissism does. We devise scads of narratives comfortingly designed with the trio of a beginning, a middle, and an end, anticipated from the outset, because we pattern our doings, be it trial, date, or soccer game.

As children, we learn subtle patterns from our parents, including the texture of their senses and their emotional style. Just as we learn the alphabet and that teeth can bite—horse teeth or brother teeth—we learn the configurations of cuddling, the emotional contours of our mother's voice, the silhouette of a friend. Many of our beliefs are woven right into the brain. This includes some truths that seem obvious and universal: a sense of three-dimensional space, fear of falling, time's passage, fear of snakes and some other critters, the basics of language, and so on. Certain kinds of knowledge don't need to be learned but do need to be activated by someone near and dear, usually Mom. As many experiments have shown, it's only when infants see a parent being frightened of a spider that they, too, develop arachnophobia. We inherit a general predisposition, which needs to be triggered. If Mom is a sky diver, heights might not alarm her; if she has a passion for creepy-crawlies, her daughter might fill her pockets with centipedes. Primal fear confirmed by others locks the fear in tight.

Thought has its precincts where the cops of law and order look for anything out of place. Without a pattern, we feel helpless, and life may seem as scary as an open-backed cellar staircase with no railings. We rely on patterns, but we also cherish and admire them. Few things are as beautiful as a ripple, a spiral, or a rosette. They are visually succulent. The mind savors them. They comfort. Societies like to invent new patterns of action, rules and rituals to cushion nature's laws under some of our own. And we devise word patterns like this palindrome: Madam, I'm Adam. But when all is said and done, they reflect one of the brain's oldest and deepest needs—to fill the world with lit pathways and our lives with a design.

Let's say you notice an object across the street. Looking at it stimulates the association areas in the lower temporal lobes, which begin to quiz and classify the image, starting with the basics: Is it large or small, moving or standing still, alive or dead, blue or yellow, human or car or baby carriage? The face recognition area identifies a likely pattern. A label for it takes shape in the left temporal lobe. It's not a wolf. It's a woman. A woman with lupus. No, that was an intrusion from the language region, an association of lupine, *wolflike*, with the illness *lupus*, so named because it can produce a wolflike mask. The parietal lobe helps you focus on the woman. With the auditory association area, you decipher the sounds— human greeting or vexed grackle? Young or old voice? Happy or worried? Meanwhile, in the brain's dreamy cities and counties, asso- ciations wake and endow the vision with meaning. The limbic sys- tem, that expressionist painter, daubs the perception with emotion. It's your mother. You recognize her hair and figure. Memory fills in her name, her upbeat personality, a phone call yesterday, her saying that she had several errands to run. Brain imaging would show activity in many different regions as you weave all the information together and think what feels like a single concept: *Mom!*

Generalizing, even from concrete details, isn't always accurate, as one discovers when a prediction doesn't materialize. You were wrong. It's not your mother. The woman across the street only bears a resemblance to her. She's not waving to you, but to the person walking behind you. Generalizing works often enough to make do in a world where things disappear on their own, and we add and subtract things from our awareness on purpose. Generalize from one traumatic fact, locale, or situation to possible others? Easy as spotting the third tenor in a trio.

A passion for pattern, piano, and defeating Hitler led movie actress Hedy Lamarr to invent the beginnings of cell phones, smart bombs, and other tricks of wireless communication. Born Hedwig Eva Maria Kiesler, the Austrian film star at nineteen swam nude in the 1933 film *Ecstasy*, a delicious scandal that brought her worldwide attention. A friend of mine fondly remembers how, night after night, he and his pals would join long lines at a Uruguayan movie theater to behold her wickedness. She married a wealthy arms merchant, who socialized with Hitler and Mussolini, and she grew to hate the Nazis as well as her husband, whom she divorced. In time, she arrived in Hollywood, anglicized her name, and fell in love with the composer George Antheil. One day, listening to Antheil play piano, she heard the notes in a new way, as changing patterns, and started thinking about an antijamming radio control that might work with torpedoes. The invention they later patented used a prearranged group of constantly changing frequencies, like the notes of a song, which made messages hard to intercept or jam. These days her invention, known as spread spectrum, enables cell phones, wireless Internet access, and the military's satellite communications, which depend on elaborate patterns of frequencies that change by the microsecond, no two the same, each attuned to its host with no attention to spare.

CHAPTER 10

In the Church of the Pines

No one can stand in these solitudes unmoved, and not feel that there is more in man than the mere breath of his body.

—Charles Darwin,
The Voyage of the Beagle

People often use religious terminology when they speak of the spiritual or transcendent. Our yearning to find whole-ness as holiness, and at-one-ment as atonement, fills a need ancient and essential as air. Because English vocabulary offers few ways to describe religious events, except in churchly terms, I often resort to such words as *sacred, grace, reverence, worship, holy, sanctity,* and *benediction,* which I cherish as powerful feelings, moods, and ideas. I'm an Earth ecstatic, and my creed is simple: All life is sacred, life loves life, and we are capable of improving our behavior toward one another. As basic as that is, for me it's also tonic and deeply spiritual, glorifying the smallest life-form and embracing the most distant stars.

If we look at the vocabulary of the Indo-Europeans, hoping to sense the texture of their lives, we discover that they invented a word for holy, which meant the healthy interplay of all living things, a sense of belonging to the whole, an appreciation for the invisible. They had a verb for "to retreat in awe" and another for "speaking with the deity." Their poet, who retreated in awe, spoke with the deity, and celebrated life's holiness, was called *wekwom teks,* "the weaver of words."

Like those first humans, we fear the dark, rejoice in being alive, feel a powerful sense of awe, and ask many of the same questions: Who and what are we? Where did we come from? How should we behave? Why is life so hard? How can beings blessed with such a powerful life force die? What is death? Our vigilant brain, questing to make sense of things, can't, but insists on trying, and sometimes consoles itself with magic, miracle, and faith. Or at least a great story. In part, religious feelings return us to childhood, when we love-worshiped our parents, whom we literally looked up to, and from whom pleasure or pain descended. And we were right to think them gods. What miracles they performed. Parents could heal a hurt with a kiss, tie magic knots, produce food and playthings, master wild animals and wobbly machines.

According to the *Encyclopedia of American Religion*, by J. Gordon Melton, the United States is currently home to 2,630 religious groups, including such unusual ones as the Unitarian Universalist Pagans, Church of the God Anonymous, Nudist Christian Church of the Blessed Virgin Jesus, Kennedy Worshipers (a secret sect that deifies the late JFK), Church of Diana Queen of Heaven (deifies the late princess), twenty-two sects that believe in UFOs, twelve that offer divinity degrees through the mail, plus all the splinter groups of traditional denominations (116 Catholic, hundreds of Pentecostal, and so forth).

We harbor such feelings of faith because we inherited pragmatic inquiring brains that also feel transcendent awe. Where is religious mysticism located in the brain? If we have so much trouble identifying a single pain center, should we expect to find a sacred grove of religious neurons? Brain mapping experiments have shown that part of the temporal lobe becomes active when subjects feel a sense of mystical transcendence. The Canadian neuroscientist Michael Persinger has gone so far as to stimulate the area in nonreligious people. "Typically, people report a presence," Persinger explains. "One time we had a strobe light going and this individual experienced God visiting her. Afterwards we looked at her EEG and there was this classic spike and slow-wave seizure over the temporal lobe at the precise time of the experience—the other parts of the brain were normal." Electrodes, hallucinogens, self-flagellation, and chanting have all summoned God for some people, maybe by stimulating the same shrine of neurons, maybe not.

Our sensitivity to religion means that at some stage in human history, our chance of mating and surviving increased if we felt a powerful sense of belonging to the pervasive mystery of nature, of being finite in the face of the infinite, of feeling bent by powerful and unseen forces, which we worshiped. Spirituality helped us survive. However, belief in a supernatural being wasn't necessary, and thus religion has taken many forms around the world, from pantheism to Buddhism to communism to extraterrestrial-centered cults. There's much to be said about this rich and distinctive vein of our human nature, but because I explore it at length in my book *Deep Play*, I won't spend longer with it here.

CHAPTER 11

Einstein's Brain

The most beautiful experience we can have is the mysterious. It is the fundamental emotion which stands at the
cradle of true art and true science.

> —Albert Einstein,
> "The World As I See It"

Some eponymous people donate their names to favorite things or
activities—Jules Leotard, Judge Lynch, and the Earl of Sand-
wich come to mind. But few become the symbol of what a human
being can achieve, both pinnacle and freak, stationed at one end of
the spectrum whose width defines the average. "Name a genius,"
you might ask people, and it's likely they'll say Einstein. Even
though he's been dead for almost fifty years, his image still adorns
T-shirts and tankards. His status as folk hero abides, despite how
few people understand his quest. It's enough that he symbolizes
genius, giving it a farouche human face surrounded by electric hair.
Naturally, scientists have wondered why he was so blessed. Einstein
did, too, and although he chose cremation for his body, he
bequeathed his brain to science.

But where *was* his brain? When Einstein died suddenly at
seventy-six, of a ruptured aneurysm of the abdominal aorta, his
brain was in remarkably good condition. Within seven hours of
Einstein's death, the Princeton pathologist Thomas Harvey had
removed it from the skull. Then Harvey took the brain to quietly
explore by himself, and had it carefully cut into 240 blocks for
study, without, alas, divining anything new. He sent tissue slides to

several neuropathologists, who reported nothing exotic, and at that point he let Einstein's brain rest for a spell.

It stayed with him, incommunicado for years, no more enlightening than the brain of an average cowhand, ballerina, masseur, or pathologist. Nothing that suggested the ability to slip around in space and time. Instead of mastering its secrets, Harvey became the brain's keeper, a combination curator and warden. For many years, the hostage floated in two mason jars of formaldehyde, inside a cardboard box marked Costa Cider, under a beer cooler in Harvey's office.

I can only imagine the guilty sense of privilege Harvey must have felt, housing the most famous brain of his age, informally, sloppily even. Did he sometimes peer into the jars and turn them gently like snow globes, talk to the brain, commune with it? Was it merely a souvenir? Was his motive to guard it from the desecration of probing fingers? Or did he entertain dreams of glory, of solving its mysteries? As the years passed, explaining how the now fragmented brain got into a cider box in his office became much more awkward.

Ultimately word seeped out, thanks in part to the prying of a New Jersey newspaper reporter, and in the 1980s Harvey permitted several scientists to study the brain. They concluded that it had more glial cells per neuron than other brains, a greater "metabolic need," which might spark a more energetic mind. That stirred public interest for a while, with many of us picturing his glia as a sort of golden mucilage, the pith of brilliance. But a closer look at the study revealed flaws, and in time it was discounted.

Then, in 1999, Sandra F. Witelson and her colleagues at McMaster University finally snared a chance to examine the brain thoroughly. What they discovered touched many nerves. It surprised and disappointed, puzzled and reassured. At 1,230 grams, Einstein's brain weighed a little less than average. It didn't boast more glial cells or neurons. In most ways it seemed perfectly ordinary. But the parietal lobes, vital for mathematical and spatial reasoning, and movement, among other things, appeared 15 percent wider than those of most other brains. Witelson had seen that sort of enlargement elsewhere, in the brains of the mathematician Gauss and the physicist Siljestrom.

Einstein's brain did seem different: the Sylvian fissure, a fold running through the parietal lobes, was missing. It's said that without that divide, neurons might find it easier to connect and communicate. Einstein had said that his mental dynamo didn't involve words, that he thought in images and solved problems in the language of mathematics. Indeed, he didn't learn to speak until he was three, and by sixteen he had taught himself calculus. Did his cunning spring from an anatomical mistake that allowed better wiring? Was it an evolutionary fluke, what some might call a flaw? Or was it more complicated than that, created from the chemical pond of his brain, a wealth of unique experiences, and the zeitgeist of the era? Did he arrive in this world hardwired for just the sort of genius mathematics requires? If so, was his calling inevitable? Or did it evolve in response to his childhood, bygone days he could retrieve from time and space with relative ease and review in his mind's eye?

CHAPTER 12

The Mind's Eye

"In my mind's eye, Horatio."
—Shakespeare,
Hamlet, I, ii, 185

Where does the moon go when one sleeps? How heavy is starlight on the eyelids of morning? If I picture a sandman dusting the corners of my closed eyes with grit, will I produce the grit to honor the reality of the sandman, because otherwise the Eden of childhood will have lost some of its myth and thunder? Sandmen toil in the memory vaults, and that's how I'm picturing them now, on tiptoe, bags of sand and silky brushes in their pockets. I think I'm borrowing them from *Winken, Blinken, and Nod,* a little golden-edged picture book I read when I was a child. I can see the book in my mind's eye, and visualize the eyelids of morning there, too. But where do those pictures hang?

I'm picturing the brain as a lightning-filled jar, where neurons fire millions of electrical bursts each moment, a silent crackling, while potent chemicals flow into and out of each neuron. Molecular messengers speed between the cells, updating the latest news. Through that lightning storm, the body speaks to the brain. Using a bazaar of molecular tongues, the mind talks to itself. In that night-filled pharmacy, bracers are crafted, analgesics dispensed, standing orders shipped to muscles and organs. Urgent communiqués are dispatched to injuries and infections. Gate-crashers race to the suburbs of a mood.

I chose tongues, lightning, bazaar, pharmacy, gate-crashers, and such because they're images familiar to my life and time, but I could just as easily have picked river delta, fireworks, railway station, shrine, crisis workers. We can talk about the brain generally if we like—I might have said *neurohormones*. We can see a brain floating in a mason jar. But we can *picture* the brain in lavish images and startling symbols. How remarkable that we can "see" something we're not actually looking at. Most often we use the mind's eye to fish out a memory, flesh out verbal descriptions, doodle, fantasize, reason, rehearse, practice a learned skill, predict consequences. We can imagine the invisible, the impossible, the unspeakable. Sometimes we furnish the mind's eye with things perceived, and at others with things that don't exist.

The brain converts the outside world into the minimum it requires: a sense of three physical dimensions plus time. It can convey several more dimensions, as Escher does in his visually paradoxical paintings, and it can think abstractly about still more dimensions, as physicists sometimes do, using mathematical symbols as mental handrails. Just as a two-dimensional painting or photograph conjures up a four-dimensional world, the brain envisions itself in space and time. But only as much space and time as it needs to survive, given its senses and limits. I often wonder about the senses of life-forms in other solar systems, how many dimensions their universe might seem to hold, depending on their biology and culture. If they have a culture, and an awareness of the outside world, and if they value truth.

We cannot know. Faith eases that strain. Faith in most anything, but especially in religion, science, and love, because they're so good at providing useful and pleasing patterns and rewards. Certainty feels sweet. Especially the certainty of knowing who and what goes where in a chaotic world. In the mind's eye, that ancient seat of imagining, neurons appear to branch like trees, and angels have taffeta bird wings.

How does the brain operate the mind's eye? Brain scans reveal that when the mind's eye is open, there's more blood flow to visual parts of the occipital, parietal, and temporal lobes—brain parts active in ordinary vision. The mind's eye uses some of the same systems and echoes the effect of really seeing something. One can inspect a mental image, say the left bank of Six Mile Creek, and focus on a great blue heron fishing almost invisibly at the far shore. The more complicated the scene, the longer it takes to picture, since we add elements one at a time, painting them on. "The reason more complex images require time to generate is simple," Stephen Kosslyn and Olivier Koenig explain in *Wet Mind.* "Each part is imaged individually, and it takes time to look up the spatial relation, adjust the attention window to the appropriate location, and form an image of the part." So, though it seems to pop into mind, hocus-pocus, the image appears in stages. The mind's eye is limited in scope. It can only hold so much detail, and images don't stay there long. Remember your mother's face. The image fades. Recall it again. It fades again. Recall it again. Now zoom in on the chin. The rest fades. Then the chin fades. Things evaporate in the mind's eye. They can be renewed, but they'll fade again. It's only a temporary arena.

In Kosslyn's lab, I took part in a fascinating MRI experiment. Wearing a stiff plastic mask that made me feel a little like an armadillo, I lay with my head in the business part of the machine, and either looked at pictures of various objects or simply thought about them, conjuring them in my mind's eye. Both activities engaged the same parts of the brain. This suggests that whether we imagine something richly in the mind's eye, or actually see it, to the brain it's nearly the same experience. Think what this implies about witnessing violent crime, remembering childhood trauma, watching violence on television, or preparing for a sports event. On one level, we know it's not real. Kosslyn finds that when most people "imagine practicing something, they see themselves from someone else's point of view." It's a form of self-mimicry or imitation. One abstracts oneself to teach oneself, and it works. Because we use the same representations to direct what we do and what we imagine we do, "refining these . . . in imagery will transfer to actual movements."

In the mind's eye, neurobiology and art meet. When an artist imagines something, to the artist it is as real as rain. Then *we* look at that landscape, hear that music, read that book, or watch that film, and on some cognitive level understand that what we're seeing isn't real, that it's an invention. But the brain processes the information as if it *were* real, and so we're partly transported to the world the artist lives in, we feel that emotion, we enter that landscape.

Human beings are sloshing sacks of chemicals on the move. The skin provides a clear boundary between self and world. Or so it seems. Since we don't have microscopic vision, we can't see our ragged edges swapping molecules with the invisibly teeming air. Thus we imagine containers everywhere, outside and in. In the mind's eye, we imagine the body full of vessels, pockets, chambers, receptacles. The blood circulates inside veins, arteries, and capillaries. Cells and plasma fill up the blood. A cage of ribs holds the heart and lungs. Muscles, tendons, and ligaments are girdled by spandex-like skin. I could have started smaller: molecules containing atoms containing quarks. I could go quantum and add that bits disappear when I look at them and can appear in two places at once. That takes some getting used to, it beggars belief, because we inherited a different understanding of the world's physics, a practical one that only works on a larger scale. Quantum physics goes against our instincts. Still, we can picture it in the mind's eye.

One of the ultimate mind's eyesights of our age is imagining the surface of another planet. It still amazes me that, aided by abstract data and skimpy pictures, our small brains can house the reality of a huge planet. I've been to a few of NASA's flybys, and in my mind's eye at the moment, I'm seeing one of them clearly: when *Voyager* flew its closest to Uranus in 1987.

In the neatly landscaped courtyard of Jet Propulsion Laboratory in Pasadena, California, the sun was thick yellow pouring onto the flower beds that day. Inside, three young NASA workers struggled to keep the sun out of the room, because it was casting glare on a movie screen. Gushing through a large window, it seemed uncontainable, a silent yellow Niagara they tried to block out with cardboard and felt and, finally, scraps of green plastic garbage bags. *Voyager* had flown 4 million miles, arrived only two minutes late, and run tests of extraordinary delicacy and sophistication, but when it came to fundamentals like cutting glare, NASA was all thumbs, all improvisation, all scraps of green plastic. In a sense, designing a spacecraft was easier—how do you hold back the sun?

Hanging from the ceiling at intervals of twenty feet, three television screens murmured about *Voyager*'s encounter with Uranus, their glowing eyes wide open. Below them a sea of rapt observers milled about, clutching Danish pastries and cups of coffee, talking solemnly or cracking jokes, pointing out celebrities or trying not to be noticed, greeting old friends, stealing glances at one name tag after another to see if there was anyone *of interest* to bump into, taking flash pictures of the TV monitors' images, but most of all, only and forever waiting. Waiting for the first thrilling pictures to roll in from Uranus and its entourage of moons.

We had gathered there with a longing to behold nature at her most mysterious and remote, to enter with the ignorance of centuries and within hours be overwhelmed by knowledge. Most people were drawn from far-flung haunts, jobs, and habits to those sterile rooms, because they felt such intense stand-up-and-salute wonder. Wonder about our cosmic neighbors and ourselves, how our molecules came to be forged in some early chaos of the sun, and what the future of our species and planet might hold. Flybys took many hours and all-night vigils, but no one wanted to sleep. Sleep and you might miss the first vibrant photograph of Olympus Mons, a large volcano on Mars; or the first shots of Titan, one of Saturn's moons, with an atmosphere a little like that of ancient Earth's; or the stark, crazed, frozen landscape of Uranus.

Few things on Earth are as thrilling as watching the first glimpses of a planet roll in. Computer-fed, they unravel down the viewing screen, row by row, as if a giant were undoing the yarns of ignorance. Like me, many people had been waiting for those startling pictures for more than a decade, from the moment *Voyager* was launched with its mission of revealing the outer reaches of the solar system. Others were there to stretch their mind's eye a little wider, and some I suppose just for the sheer spectacle. Every encounter was exhilarating for different reasons. The amazing moons of Jupiter no one had seen up close before. The rings of Saturn that looked like ribbons of sherbet. The gauzy white face of Venus' acid clouds. The unthinkably barren and remote moons of Uranus. The windy, red geology of Mars. Where else could one glimpse the same deserts and seas people have ached to visit for as long as there have been people?

Even though I'm sitting in my study today, watching an autumn snowfall quilt the garden in downy white, and that Uranus encounter feels light-years away, I can easily conjure it up. "Men go out and gaze in astonishment at high mountains," wrote St. Augustine ages ago, "the huge waters of the sea, the broad reaches of rivers, the ocean that encircles the world, or the stars in their courses. But they pay no attention to themselves. They do not marvel at the thought that while I have been mentioning all these things, I have not been looking at them with my eyes, and that I could not even speak of mountains or waves, rivers or stars, which are things that I have seen, or of the ocean, which I know only on the evidence of others, unless I could see them in my mind's eye, in memory."

PAVILIONS
OF DESIRE

(Memory)

CHAPTER 13

What Is a Memory?

What sort of future is coming up from behind I don't really know. But the past, spread out ahead, dominates everything in sight.
—Robert M. Pirsig,
Zen and the Art of Motorcycle Maintenance

L ike tiny islands on the horizon, they can vanish in rough seas. Even in calm weather, their coral gradually erodes, pickled by salt and heat. Yet they form the shoals of a life. Some offer safe lagoons and murmuring trees. Others crawl with pirates and reptiles. Together, they connect a self with the mainland and society. Plot their trail and a mercurial past becomes visible.

Memories feel geological in their repose, solid and true, the bedrock of consciousness. They may include knowing that it's hard to lead a cow down steps, or how the indri-indri of Madagascar got its name, or the time you accidentally grabbed a strange man's hand in a crowd (thinking it was your friend's), or how you felt hitting a home run in Little League, or your first car (a used VW that rattled like an old dinette set), or a grisly murder you just read about that made you rethink capital punishment, or an unconscious detailed operating guide to the body that manages each cell's tiny factory.

Memories inform our actions, keep us company, and give us our noisy, ever-chattering sense of self. Because we're moody giants, every day we subtly revise who we think we are. Part of the android's tragedy in the Ridley Scott film *Blade Runner* is that he possesses a long, self-defining chain of memories. Though ruthless and lacking empathy, and technically not a person, he can remember. Played by Rutger Hauer, he contains a self who witnessed marvels on Earth and Mars and fears losing his unique mental jazz in death.

Without memories we wouldn't know who we are, how we once were, who we'd like to be in the memorable future. We are the sum of our memories. They provide a continuous private sense of one's self. Change your memory and you change your identity. Then shouldn't we try to bank good memories, ones that will define us as we wish to be? I'm surprised by how many people do just that. Even tour companies advertise: "Bring home wonderful memories." *Here we are, a happy family taking a Disney cruise,* documented on film. But memory isn't like a camcorder, computer, or storage bin. It's more restless, more creative, and it's not one of anything. Each memory is a plural event, an ensemble of synchronized neurons, some side by side, others relatively far apart.

Everyone will always remember where they were on September 11, 2001, or when men first walked on the moon. Shared memories bind us to loved ones, neighbors, our contemporaries. The sort of memory I'm talking about now isn't essential for survival, and yet it pleases us, it enriches everyday life. So couples relive romantic memories, families watch home movies, and friends "catch up" with each other, as if they've lagged behind on a trail. Sifting memory for saliences to report, they reveal how vital pieces of their identity have changed. Aging, we tailor memories to fit our evolving silhouette, and as life's vocabulary changes, memories change to fathom the new order. Lose your memory, and you may drift in an alien world.

Mind you, memories are kidnappable. Radio, television, and the print media purvey shared national memories that can usurp a personal past. All the why's can change. A world of artificial memory, as the British neuroscientist Steven Rose points out, "means that whereas all living species have a past, only humans have a history." And, at that, it tends to be the history of the well to do. Thanks to the compound eye of the media, millions of people are spoon-fed the same images, slogans, history, myths. What happens to individual memories then? Some rebels refuse that programming, or they prefer their own group's ideologies. But most people do adopt values and interpretations of events from the media, their neighbors, or a favorite tyrant. Official history changes with each era's values, which can sometimes be perverse, what Jung described as a large-scale psychic ailment. "An epoch," he said, in *Modern Man in Search of a Soul*, "is like an individual; it has its own limitations of conscious outlook, and therefore requires a compensatory adjustment ... that which everyone blindly craves and expects—whether this attainment results in good or evil, the healing of an epoch or its destruction." Still, though no one is an island, most are peninsulas. Our lives wouldn't make sense without personal memories pinned like butterflies against the velvet backdrop of social history.

Scientists sometimes talk about "flashbulb" memories so intense they instantly brand the mind. Photography provided something different: push-button memories that revolutionized our sense of self and family, which we often remember in eye-gulps, as snapshots. Walt Whitman, in his journals, jotted down the name of each of his lovers and sometimes what they did for a living, as though he might one day forget his moments of loving and being loved. But I think he would have preferred photographs of those dear ones to help recall the liquid mosaic of each face.

Picture yourself younger, and what image forms? Most likely it's a static image, a snapshot someone took. Memories can pile up and become mind clutter; it's easier to store them in albums. We remember our poses. Each photograph is a magic lamp rubbed by the mind. When we're in the mood, we can savor a photograph while sensations burst free. Right now, for example, I'm holding a photograph of a pungent king penguin rookery in Antarctica, and I remember the noisy clamor like a combination of harmonica and oncoming train. I remember how inhaling glacial cold felt like pulling a scarf through my nostrils. I remember that, in such thin air, glare became a color.

Whenever we look at a photo, we add nuances, and that inevitably edits it. It may pale. It may acquire a thick lacquer of emotion. The next sentence may sound a little bizarre because

English grammar isn't congenial to time mirages, however: pho-tographs tell us who we now think we once were. Photography, like most art, stores moments of heightened emotion and awareness like small pieces of neutron star. Years later, a memory's color-rodeo may have faded, or may remain vivid enough to make the pulse buck again. Each response adds another layer until the memory is encrusted with new feelings, below which the original event evap-orates. Imagine a jeweled knife. First you change the handle, then you change the blade. Is it the same knife?

We tend to think of memories as monuments we once forged and may find intact beneath the weedy growth of years. But, in a real sense, memories are tied to and describe the present. Formed in an idiosyncratic way when they happened, they're also true to the moment of recall, including how you feel, all you've experienced, and new values, passions, and vulnerability. One never steps into the same stream of consciousness twice. All the mischief and mayhem of a life influences how one restyles a memory.

A memory is more atmospheric than accurate, more an evolving fiction than a sacred text. And thank heavens. If rude, shameful, or brutal memories can't be expunged, they can at least be diluted. So is nothing permanent and fixed in life? By definition *life* is a fickle noun, an event in progress. Still, we cling to philosophical railings, religious icons, pillars of belief. We forget on purpose that Earth is rolling at 1,000 miles an hour, and, at the same time, falling ellip-tically around our sun, while the sun is swinging through the Milky Way, and the Milky Way migrating along with countless other galaxies in a universe about 13.7 billion years old. An event is such a little piece of time and space, leaving only a mindglow behind like the tail of a shooting star. For lack of a better word, we call that scintillation *memory*.

CHAPTER 14

Reflections in a Gazing Ball

The Brain—is wider than the Sky—
For—put them side by side—
The one the other will contain
With ease—and You—beside.
    —Emily Dickinson,
        "The Brain is wider
        than the Sky"

Memory is the brain's main job, and a favorite arena of research. Ironically, that effort enlists the memories of many humans, animals, and computers, connecting them in a weave stretching back to the origins of life on Earth, and it includes the entire animal kingdom. Only by holding a mirror and a lamp can we peer into our hive of memories, some tidy as geometry, others sloppy as tapioca. A key problem is memory's several faces and abodes. It inhabits various locales in the brain, funds most of the brain's enterprises. "Thanks for the memory," Bob Hope's signature song goes, emphasizing the succulent gift of memories, even bittersweet ones. Here's a bittersweet one of my own:

One fall morning when I was six, I hurried through an orchard with three schoolmates. We were late for first grade, and there were going to be silhouette drawings none of us wanted to miss. I remember the shiny plaid dress that Susan Green wore, her matching hair ribbon, and a petticoat that rustled as she moved. Ripening apples spiced the air with scent. High in the branches, dark plums huddled like bats. Susan dragged at my arm because I'd slowed to stare at the plums, her eyes followed mine, and when she demanded to know what I was looking at, I told her. Suddenly she let go of my arm and all three girls recoiled. The possibility of bats didn't frighten them. *I* frightened them. I looked at plums and saw bats. The alarm on their faces became an indelible memory, one colored by shame, and fused with a nearly levitating sense of wonder.

That vignette holds the kernel of many truths about memory. To remember, the brain does four things superbly: recognizes patterns, interprets them, records their source, and retrieves them. In the orchard, I saw abdominal apples, smelled their corklike sweetness, heard the warble of children, held unforeseen feelings. At the time, brain receptors combined all the stimuli fast, busily interpreting what an *orchard* meant, the unique scent of apples, the similar architecture of plums and huddled bats, and noting the whereabouts as a walk to school, while it piled up emotions.

Now, half a lifetime later, only shards of that orchard memory float into my ken—light swerving off a satin skirt, clouds caged in the branches of a tree, the hug of an elastic waistband on the corduroy pants I wore under my dress, the scratchy whispers of Susan's starched petticoat. But I can forcibly capture others. Not the other girls' names, not the cadence of their voices. Lots of details have faded through the normal vanishing and crumbling known as graceful degradation. Like an old photograph, my recall of that day has lost some of its color and clarity with passing years. But, unlike a photograph, the memory isn't stored whole. It's distributed throughout the brain and slowly dissolves, sheen by warble, whiff by shame.

Fortunately, memories nest in elaborate thickets of association. If I concentrate hard and insinuate myself back into that moment, into that body, peering out behind curly bangs, I can feel the swing of my ponytail, and my body flushed with wonder the way a cube of sugar absorbs water. The sort of sugar cube with a slightly bitter pink splotch we received in squat, pleated white paper cups one day at school when the first polio vaccines were distributed.

My memory caroms off another association. The sort of sugar cubes a stable-hand placed onto my trembling palm one summer, and a horse gummed off gently with velvety lips that tickled. Not just the horse's lips but the whole experience tickled, and I laughed, which didn't frighten the horse, used to the antics of pint-size humans. Especially if they made high-pitched noises. Had I hissed or growled, the horse might have bolted. But in *its* long-term memory, it knew that baby horses whinnied a sprightly coloratura, and maybe that the helpless young of other mammals made high-pitched noises, too. If I'd been bitten by that horse, the event would have burrowed deep into my memory and advised my reflexes as well as my conscious mind. Even deeper in the horse's long-term memory, and ours, lodges the sound of a cat's claws scraping on a rock (why chalk squeaking on a blackboard chills us?), an alarming sound, an instant warning. An animal rarely gets two chances to survive a life-threatening attack. It must store the memory faster than a snake strike, and keep it on tap for a lifetime.

Indeed, pain and memory have a lot in common. Both change NMDA receptors, allowing them to open more easily and stay open longer. Pain is like a bad memory that won't fade, which makes good evolutionary sense since it's not enough to identify danger, an organism also has to remember its features. Under general anesthesia, one may not be aware of pain, but the nervous system remembers it well. At the University of Pennsylvania Medical School, Allan Gottschalk and his colleagues have been administering "preemptive analgesia," pain medicine beforehand, not just during or after an operation. Hoping to keep chronic pain memories from forming in the first place, they block the nerve pathways leading from the injury site to the spinal cord. Thus far, it's working well in prostate operations.

Back to the orchard. The long string of facts about apples, horses, orchards, and such involve *semantic* memory. Remembering that specific morning and those girls—on a day when the plums caught my eye, the clouds seemed caged by the branches, and I felt both shame and wonder—employs *episodic* memory. Combine the two and you get *declarative* memory, a declaration of events easy to slap a word on, the explicit truths. For both storage and retrieval, such memories rely on the sea-horse-shaped hippocampus (from the Greek for "sea monster").

Unconscious memory and knowing *how* something happened involves many areas of the brain, including various sensory and motor pathways. Knowing *how* may be the record kept by my body of what it was like to walk through the orchard. These subtle skills evade words, but the body remembers itself through a wardrobe of mannerisms, habits, preferences, gestures, biases, styles of talking and thinking. Perhaps a walk that's more amble than stride. Or how, when nervous, you unconsciously pick at your cuticle. How volatile you become when provoked. How little it takes to provoke you. Such minutiae rule and define us, allowing a body to remember the individual it retains. We need to think about swimming while we're learning, but afterward the body remembers how to float, angling a hipbone just right, without consulting us.

There isn't a single place, a memory-mine, where all the ore of experience lies buried. Different types of memory inhabit different parts of the brain, and groups of neurons combine to form a single memory. A good way to picture this is offered by the neurologist Jeff Victoroff, who suggests a football stadium at halftime.

Twenty thousand people sit across from you, each holding a colored card. At a signal, they flip their cards into position, and the pattern spells out a message: "Go Trojans!" The idea "Go Trojans" is not written on any *one* of those 20,000 cards; it is not located at any one seat. It exists only as a pattern of activity, like the coordinated firing of 20,000 neuronal responses. In the same way, memories are stored in our brains not in any one *place* but as a distributed network of neurons, primed to flip their cards of synaptic activity in a coordinated way.

Continuing that metaphor, one excited person can somehow rally all the others. Stimulate one facet of a memory and the whole can suddenly pop into mind.

Locked forever inside the body's ribbed prison, we can only know life through our avid, prowling senses. They report to a region near the hippocampus where their information creates a multifaceted picture of an event. After several more steps, the information returns to the hippocampus or to the neocortex for storage. We might store facial memories in the temporal lobes, a landscape in the parietal, and socializing in the frontal lobes. But, since we remember a whole event, not a spray of sensations, everything blends in the large association cortices that make up most of the neocortex.

Somehow, all these ways of remembering combine and we feel singular. Add enough pieces to the mosaic and an individual finds shape. We take for granted these dazzling skills, and the most treasured gift of all, being able to time-travel and explore the lost kingdoms of yesterday. We may be the only animals with this rich form of episodic memory, in which we can revive our past, play it back like a film we stop to look at, enter imaginatively, and revise as we grow older.

CHAPTER 15

Remember What?

"Ah! that is always the way with you men; you believe
nothing the first time, and it is foolish enough to let
mere repetition convince you of what you consider in
itself unbelievable."

—George MacDonald, *Phantastes*

Say *memory,* and almost everyone thinks of the past. But most of
our memories are really about the future. We quarry experience
to help us solve present problems. A hungry squirrel needs to
recall nuts buried at the base of a sycamore tree. A mother lion
needs to recognize the scent of her cub. We need to remember a
swindle or bonus to foresee the outcome in a similar situation.
Overlapping beacons of memory guide an animal through an
ambiguous, confusing world, in which it runs a four-dimensional
obstacle course of threats, injuries, and challenges. It must survive
long enough to mate and rear young who will pass the spiral baton
of its genes and theirs to their offspring in a relay race that began so
long ago no one can remember who first ran it, or where.

As I write this and you read it, we're using a golden kind of short-
term memory called *working memory,* a mental draft horse. Our
working memory provides the general feel of our days: bite of
lemon, skid of linen, tang of joy when a loved one wakes, repeating
a telephone number we're dialing, rehearsing what to tell the
plumber, doodling a chickadee hanging upside down from an ice-
glazed twig—all while receiving updates from the body on the
career of one's tummy, muscles, or worry. Working memory holds

crates of information for immediate use, but it can only do one thing at a time. Interrupt someone while she's trying to remember a phone number and she'll probably lose her *train of thought*, as we like to say, as if a thought moved with locomotive force in a straight line on steel rails. Involving the frontal lobes (a little above and behind the eyebrows), working memory combines sensory news, the emotions that arouses, and our conscious effort to remember something. Or, if you like, every train of thought has many cars. In the brain, they link together before we can act on an impulse, solve a problem, talk to someone, feed ourselves, daydream. One stores facts. Another sifts through sensations and disposes of distracting ones. Yet another is devoted to learning skills. Another works the body's muscles. Others haul such essentials as language, social protocols, sensory routines, biodegradable resources, operating guides, rules for executive functions and rules for grunts, and many different grades of memory. They gyrate together in a single train of thought.

Whenever we learn something, the brain mints new connections or enlivens old ones. A few times each summer, I spruce up the mulch pathways through my garden, making them more walkable. The brain renews familiar pathways, too. I find this especially comforting in middle age, when my memory seems to be fading like an old picnic blanket, and recalling a simple word may take an annoying amount of time and effort. I can see the object in my mind's eye; it just takes longer to fetch the word for it. Not a problem when I'm writing (because it normally requires an extra millisecond to escort a word from brain to paper), it can turn a simple conversation with oneself or someone else into a maze of circumlocution. Not being able to recall the word *awning*, for example, I might retrieve it by picturing and naming its associates at speed (*porch, window, chaise, wicker*, and so on) until the word appears under the same heading, or mental awning. Or I may furiously search for a synonym among awning look-alikes (*drape, canvas, umbrella*), making do with a close word instead. I may picture the house *yawning* outward and hunt the verb. This runaround can be worrisome and frustrating. It upsets the rhythm of quick elusive chat and backchat. Thus it becomes not so much a roadblock as a hurdle (a word I just had to grope for and only found by picturing an eponymous track-and-field event). A personal myth is that such catastrophes appeared one day when the calendar began tearing off its own pages. But, of course, it always starts earlier, this angling for a sparkling word that darts away. Young children tend to recall events in lively detail, but that youthful gift already begins to pale at puberty, by which time life offers so many sensations that one can't remember them all.

It's right on the tip of my tongue, we say picturesquely, as if a moth were perching on the taste buds. Tip-of-the-tongue memory is more complicated than it seems. Remembering a word takes two steps, pinpointing the word you want and then fetching the sound code for the word. It's possible to retrieve only the first part, a semantic idea of the word, and not be able to remember its sounds, due to weak connections. Even though *awning* was hard to excavate, it did pull free. I could think and name it. We collected such memories at least once before, and now re-collect them like a basket full of mushrooms.

Without the brain's temporal lobe, conscious recollection wouldn't happen. That's why Alzheimer's, an illness that depletes the temporal lobe, begins slowly in a plague of forgetfulness easy to confuse with the forgetfulness that's a normal bane of aging. In time, the disease invades other parts of the brain, attacking all forms of memory, until a recognizable self—one of memory's better pranks—disappears like chalk underneath the brisk strokes of a blackboard eraser. "I am not in my perfect mind," King Lear laments. "Methinks I should know you, and know this man; yet I am doubtful; for I am mainly ignorant what place this is, and all the skill I have remembers not these garments; nor I know not where I did lodge last night."

The neurologist Spencer Nadler often receives e-mails from his patient Morris, who is in the early stages of Alzheimer's and, miraculously it seems, can eloquently describe his loss of mind while it's deteriorating. "Thoughts no longer percolate in my brain," he confides. "They've slowed, become viscous." Having the illness is

not worse than knowing he has it. "Living with incurable, pro-
gressive dementia, the horrors of the fact and the illness combined,"
he writes to Nadler, "is like living in the aftermath of a nuclear war,
surrounded on all sides by devastation and waiting for the radiation
sickness to make you finally wither and drop." More philosophical
and upbeat than most would deem possible in such circumstances,
he one day says in an e-mail: "What difference does a little dementia really make, when the greatest minds have struggled in vain to
know themselves and others?"

About 4.5 million people now suffer with Alzheimer's, and
that number is estimated to climb to 16 million by 2050, when the
baby boomers will enter their anecdotage. There's early-onset and
late-onset Alzheimer's, with the early version being more hereditary. Surely a magic drug called something like *abracadabra* will
unlock the rusty gates? The consensus is no, not yet, but someday
soon. Meanwhile, drugs like Aricept oil the hinges (by restoring lost
acetylcholine to the synapses), and laboratory wizards are targeting
the disease from many angles, some focusing on the twisted tangles
of tau proteins, others on the bull's-eye-shaped plaque.

Genes order the making of proteins, which are strings of amino
acids that normally can loop, curve, twist, and even form pleats,
through a million gyrations, before reaching their final form. But
their shape dictates how they'll behave in the body, so the strings
must fold correctly, and many don't. What to do with the tangled
protein strings? Cells have protocols for dealing with them, chap-
erone molecules to surround and protect them, others to tag them
with the amusingly named "ubiquitin" and drag them off to slice up
and recycle. But the janitorial system doesn't always work. "A certain amount of misfolding is fine," says Fred Cohen, of the University of California at San Francisco. "The cell can handle the
trash. But if there's a garbage strike, the trash on the sidewalk
begins to stink. That's what we're dealing with." Some scientists
believe that Parkinson's, Alzheimer's, mad cow disease, and many
other scourges are misfolded protein diseases. In Alzheimer's,
thanks to various saboteurs—such as the enzyme beta-secretase,
which clips proteins sticking out from the cells, leaving the stumps

in a muddle—clusters of deformed proteins clog up regions of the brain involved with memory, location, and mood. Why that happens isn't clear—genes? exposure to toxins? inflammation? trauma? But people with larger brains, more education, or complex thought seem to resist its symptoms longer.

What's come to be known as the Nun Study recently offered intriguing clues to Alzheimer's. Although epidemiologist David Snowdon, of the University of Kentucky, limited his study to a small homogeneous group of nuns (the School Sisters of Notre Dame), it boasted 90 percent accuracy, and suggested that one's thinking style while young can predict the disease in old age. Analyzing the nuns' writings as young women, before they took their vows, Snowdon and his colleagues found that those using the simplest sentences with the fewest ideas were the ones most likely to develop Alzheimer's later in life. What are we to make of this? Maybe just that people with lots of cell connections fare better because they can afford to lose more. Or does nutrition play a role? The Nun Study also discovered that the cerebral cortex weakened in sisters with low levels of folic acid.

Although the brain ages dramatically like the rest of the body, old age doesn't guarantee senility. Retrieving a particular word may take longer, but vocabulary can actually improve. There's lots of evidence that seasoned skills age more slowly. Habits of mind linger, especially expertise. Anton Bruckner continued composing lush melodic symphonies into his seventies and eighties. I'm always pleased to see nonagenarian physicist Hans Bethe (among a great many accomplishments, he figured out how the sun works) strolling in the Cornell arboretum with his wife, friends, and colleagues. At ninety, he signed a five-year book contract, and he continues to stay politically and professionally active. As the most senior living scientist who worked on the atomic bomb at Los Alamos in the 1940s, he's been campaigning ever since for nuclear disarmament. Much of each day he happily devotes to the study of exploding stars. Yet other densely imaginative thinkers—Iris Murdoch comes sadly to mind—have endured the mental deforestation of Alzheimer's. Both Bethe and Murdoch spent their lives juggling complex sentences and ideas.

At the local airport ten years ago, I happened to be in line behind Bethe (pronounced BAY-ta) at the ticket counter and overheard the clerk say to him, slowly and in a louder voice than needed, as if he were carrying an invisible ear trumpet and must, at his age, be lost in senility:

"Now, Mr. *Beth-ee,* you'll be arriving at gate 21 in Pittsburgh and going to gate number 27. That's *six* gates away."

A small bemused smile flitted across Bethe's face. "Oh, I think I can do the math," he said.

Most everyone I know frets about memory loss, and, in what's become a mass phobia, worries whether each slip foretells the reign of Alzheimer's. Maybe because it can be so embarrassing, people seem especially bothered when they forget familiar names. I know several couples who have chosen a help-I-can't-remember-the-name signal; the other quickly introduces him/herself and hopes the person the couple has just bumped into will do the same.

We complain about normal forgetfulness, but thank goodness we don't have better memories. We aren't required to remember how to operate our bodies, for instance, or even the full carnival of sense impressions spawned by a single moment. Remember every slight and insecurity since childhood? People cursed with comprehensive memories have minds like overstuffed closets—open the door and an avalanche pours out. A simple choice can balloon into a nightmare of competing outcomes. Not a good survival plan. Forgetting isn't the absence of remembering, it's memory's ally, a device that allows the brain to stay agile and engaged.

Running a peak-performance animal is expensive, so the body devotes only the first half of life to high-octane survival and breeding skills, which include salting away lifesaving memories. After childbearing, memory begins to deteriorate, as does the rest of the body—joints grow less supple, taste buds less sensitive. In the evolutionary scheme of things, infertile humans are socially helpful but not essential, and they do compete with the young for food and shelter. But we're willful beasts who don't always play by evolution's rules. We insist on learning them, though. Part of our great charm as a species is our passion to understand everything that touches our lives. Solving mysteries is the brain's fetish and pastime, offering a special caliber of pleasure that falls between thrill and relief. We do it for survival, we do it for work, we do it for play. Fortunately, nature holds more secrets than we can unpuzzle.

But some are yielding. One mystery that's tantalized people for ages is how the brain embeds long-term memories. As I mentioned earlier, brain cells communicate by sort of shaking hands at hundreds of billions of minute contact points called synapses, slender inlets between neurons. To store a long-term memory, a cell lathers its handlike axon with more protein, which strengthens the grip. It may enlarge synapses or create doubles or triples of the synapses already there. But it's an elaborate process, and for good reason. Short-term memories may swarm and vanish, but if we clutter our minds with them for long, we drown in a sea of formed and forming memories. To prevent that, installing a long-term memory requires the safety lock of simultaneously switching on some genes while switching off others. The brain checks and double-checks. Eric Kandel, of Columbia University, theorizes that age-related memory loss might have to do with a defective genetic switch, so that short-term memory doesn't convert to long-term with a spurt of new proteins. And some people who seem to have an extraordinary memory may simply have a defective part of the switch that inhibits genes. In the mansions of their mind, lamps are mistakenly left on because the switches are broken, so they continue casting light.

In 1949, the Canadian psychologist Donald O. Hebb proposed that memory stems from neurons working in unison to strengthen the synapse where they meet. We're sociable beings, even on the cellular level where active neurons cement their mutual bonds, forming "little cliques, or social clubs, within the brain," as the neurosurgeon Frank Vertosick puts it. They really *are* social clubs, societies of cells, and some will be influenced more than others by lobbying neurons. Just as in a human society, the majority reaches a decision, despite naysayers. But altruistic neurons don't act sensibly for the good of their little clique. They act separately, selfishly, to promote their own genes, oblivious to the others, and it doesn't matter a whit if some aren't on the bandwagon.

I prefer to think of the Hebb rule as a bedroom drama. The more often a neuron excites another neuron the easier that becomes. The more two neurons excite one another, the tighter their bond grows. The reverse is also true: a neuron inhibiting another weakens their bond. Neurons love to be turned on, to feel alive, and they prefer exciting contacts over those that turn them off. A neuron doesn't turn on instantly, it requires a little buildup until a threshold is reached. Once aroused, it fires briefly down its length, is drained of its juice, then must rest a while and recharge before it can perform again. If a partner is still very excited, it may respond by firing again and again; if not, it may quiet down. In neuroscience parlance, neurons that fire together wire together.

Extending this idea in 1973, Timothy V. P. Bliss and Terje Lømo of the University of Oslo discovered that if they stimulated nerve cells in the hippocampus with high-frequency electric pulses, the cells became tightly linked. That stronger grip, known as long-term potentiation, can hold a memory in readiness for hours or weeks. Later studies by other scientists revealed that applying low-frequency pulses to hippocampal pathways weakened the connections. So, as I've said, many believe this two-step of strengthening and weakening of synapses allows various precincts of the brain to store and erase information of various sorts. In the lima-bean-size amygdala, for instance, memory is most likely involved with emotion, since part of the amygdala's role is to link emotions and experience.

In the 1980s and 1990s, several researchers identified doughnut-shaped molecules (NMDA receptors) as key to the cell changes. Seated in the outer wall of certain neurons, they're essentially coincidence detectors. Their central doorway is closed with a double lock. If transmitting and receiving neurons fire simultaneously, the door opens and current flows in, helping the brain associate two events. Timing is everything. In older animals, two signals must arrive almost simultaneously for the cell to open its gate, but in younger ones signals can be relatively far apart (a tenth of a second), which may explain the rich and enduring memories of the young, while adults find it hard to learn new things. It's tempting to imagine two signals colliding on the doorstep—a biophysical open sesame—and then, door agape, memory pouring in like cement.

But as Larry Squire and Eric Kandel remind us in their overview of the field, *Memory: From Mind to Molecules,* a memory isn't instantly engraved, but takes time and several steps to embed:

> Until the process is fully completed, memory remains vulnerable to disruption. Much of this process is completed during the first few hours after learning. But the process of stabilizing memory extends well beyond this point and involves continuous changes in the organization of long-term memory itself.

As time passes, the hippocampus plays a lesser role and memories gradually join others in different parts of the brain, forming strata of belief about the world and oneself. The first time you hear a soprano singing the beautiful lyric "The Sun, Whose Rays" from *The Mikado,* you may not recognize it. But after being repeatedly rewarded for associating those sounds with Gilbert and Sullivan—rather than with wren's song or jackhammer—some synapses find it familiar. Their doors quickly open and you think, *Mikado.*

Not long ago, the molecular biologist Joe Z. Tsien of Princeton University doctored the genes of mice to see if he could improve their memories. First he bred mice without a vital part of the NMDA receptors in the hippocampus. Those mice had weaker synaptic connections and poorer memories. Then he reversed the experiment by enlivening the receptors. Sure enough, it created brainier mice, which he named Doogie, after the young genius on the TV sitcom *Doogie Howser, M.D.* The receptor doors stayed open only 150 thousandths of a second longer than usual, but that brief welcome boosted memory and intelligence fivefold. Suddenly the mice could learn all sorts of things faster—emotional, spatial, and auditory—and retain them longer.

On the way outside, I leave the screen door open an extra slice of a second. It seems simple and fast, yet it's long enough for one of the resident hummingbirds to fly into the house and make mischief, or a june bug to clatter around the living room like a windup toy, or a triangular stinkbug to sneak in like a delta-wing bomber. We're used to gross amounts of time, like stores remaining open half an hour longer, or pearl divers holding their breath an extra minute, or locking eyes with someone for whole seconds. In one second a tuning fork set to A above middle C will vibrate forty times, a human heart will beat once, a moonbeam will almost reach the Earth, Americans will eat 350 slices of pizza. An average housefly flaps its wings every three thousandths of a second. So, 150 thousandths of a second is hard to imagine, let alone its lifelong effect on someone's destiny. One unstated innuendo is that animals *can* be smarter, that they're not fully evolved yet. If that's true of mice, it's probably true of humans. In future days, will our descendants regard us as primitive lowbrows much as we regard the Cro-Magnons?

Tsien is often asked if we can engineer smarter children now, produce a tribe of geniuses. "The short answer is no," he says, "and would we even want to?" Many parents would. But intelligence hinges on an assembly of talents, not just good memory, dexterity, or a knack for problem solving. Different animals need to be smart in different ways. Whales, pigeons, and butterflies are astonishing navigators; dogs can detect a person's essence in his footprints, picking up scent molecules that have seeped through shoe soles. But boosting their IQ wouldn't make them good at nuclear physics or the tango. In a way, one-celled organisms may have a truer sense of the world, because they respond to every stimulus they encounter, whereas higher animals like us select very few stimuli from the vast array.

In any case, a better memory isn't a more useful memory. Studies show that the IQ range of most creative people is surprisingly narrow, around 120 to 130. Higher IQs can perform certain kinds of tasks better—logic, feats of memory, and so on. But if the IQ is much higher or lower than that, the window of creativity closes. Nonetheless, for some reason we believe more is better, so people yearn for tip-top IQs, and that calls for bigger memories. A fast, retentive memory is handy, but no skeleton key to survival. I'd add "or to happiness," but survival doesn't guarantee it any more than our founders did in the Constitution. All those gents promised was the right to *pursue* happiness.*

Because IQ tests favor memory skills and logic, overlooking artistic creativity, insight, resiliency, emotional reserves, sensory gifts, and life experience, they can't really predict success, let alone satisfaction. Some theorists suggest other ways to portray intelligence. For instance, Daniel Goleman writes persuasively about our powerful emotional intelligence that colors all our decisions. An agile memory doesn't guarantee balanced intelligence (consider idiots savants), but one can't be very intelligent without a good memory.

Suppose I don't want to doctor my genes? Repeating something helps to store it long term. The dreaded tedium of rote learning, whose epitome I suppose is having to confess your bad behavior a hundred times on the blackboard, works tiresomely well. Athletes and surgeons alike develop mastery through repetition, a popular and reliable way to remember. On the other hand, tedium bars richer possibilities. Rote learning works well, but it's rigid. It doesn't allow creative solutions to bubble up through the magma on the spur of the moment. When a skill becomes automatic, the swampy path that led to it disappears, and it's hard to make adjustments that might improve one's gait or route. A theater director I know advises his actors to prepare lavishly through study and practice, then put habitual thought aside, risk spontaneity, and see what happens.

I use a skill first devised by the Greek poet Simonides in 500 B.C. Noticing the brain's natural tendency to fish for memories by way of associations, he would decorate a person's name with a colorful image. Attaching a clever hook to the slippery name makes it easier to grab and haul into view. I may choose rhyming slang, a ghoulish cartoon of one feature, a clue based on personality. These aren't always delicate or civilized tags, so I don't tell people about them. Using such a memory aid means adding an extra scintilla of time between saying "Hi" and adding "Jane." But it usually works, and it's fun, if you remember not to giggle when you say "Hello,

---

* I discuss that elusive goal a little later, in "Never a Dull Torment."

Lester," while thinking he's swollen like a blister, or Ginny likes to drink gin at lunch. If Brenda is a redhead, I might picture her hair on fire, because I know the word Brenda, cognate with the German *brennen*, hides the idea of burning at its root, or roots. Holly I might remember by her halo of curly hair. Paul's broad shoulders might remind me of a wall. One semester I began a course by confiding this technique, and we went around the room confecting funny images for each name. For once, everyone knew everyone else's name from the start. Imagination's muscle grows stronger with use, but toning memory calls for the ingenuity of Simonides. Memory techniques, though fun at times, take work, and we lazy sorts hate to bother. Anyway, the sweetest memories arrive unbidden, as gifts given, like flying fish leaping into the sunlight for a moment before plunging back into the dark, churning sea of the unconscious.

"Don't do an all-nighter before a test" is good advice, since many studies show that sleep improves recall. Sleep may aid memory through a process known as *interleaving*—gradually introducing the information through repetition, rather than just dropping it in as a chunk the brain has to assimilate. In this way, folded like egg whites into the batter, the new information doesn't disturb the old, it simply adds to it.

Cumin, an anti-inflammatory spice central to Indian cooking, may help memory. Traditional herbalists have always used melissa, also known as lemon balm, to improve memory. Lemon balm is a lovely garden herb with soft leaves and a mild lemon smell and flavor—in contrast to lemon verbena, whose scent and flavor is stronger and whose leaves feel raspy. I prefer the smell of verbena, but often add lemon balm to tea. Researchers at the Human Cognitive Neuroscience Unit at Northumbria University have isolated the herb's active ingredient, methanolic extract, which boosts the level of the neurotransmitter acetylcholine. They tried it out on volunteers; others received a placebo. Doses of about 1,600 mg produced "striking" results: a combination of calmness and improved memory.

One thing is certain: memory suffers when we're under stress. Both stress and tedium can kill brain cells. Challenge, novelty, and rich environments can rejuvenate memory. So does gentle aerobic activity, and, quite possibly, eating a cup of blueberries each day. So, if one needed an excuse to adopt a fascinating hobby, work on novel projects, or live outrageously, one could claim to be doing it for health purposes. How does this work? In the forest of neurons, transmitter chemicals leap about like capuchin monkeys. Stimulate the brain and the neurons grow new branches, which makes it easier for the monkeys to travel from one tree to the next. Fortunately, this growth process isn't limited to one's first three years but continues over a lifetime. Maybe one can't teach arthritic legs to sprint. But one can always teach an old dogma new tricks.

CHAPTER 16

Remember, I Dream

Stand there, baulked and dumb, stuttering and stammering, hissed and hooted, stand and strive, until, at last, rage draw out of thee that *dream*-power which every night shows thee is thine own; a power transcending all limit and privacy, and by virtue of which a man is the conductor of the whole river of electricity.

> —Ralph Waldo Emerson, "The Poet"

One theory about dreaming is that it evolved as both a roadside rest and a memory vat, a place to consolidate all kinds of things. It's such a grab bag of old and new, abstract and concrete, visual puns and symbols, that unlikely elements can become wadded together. So it's little wonder dreams sometimes churn up creative insights. We don't spend long each night in the boudoir of dreams, only about two hours during REM (rapid eye movement) sleep. But over a lifetime, that adds up to about six years. Why do the eyes move? The brain's inner electric continues to thrum while we sleep, but most sensory information and movements are blocked. Dreams contain so much commotion that the sleeper's body must be temporarily paralyzed or she would jump up and run around in a hallucinogenic delirium. But there's no danger if the eyes stir like moth wings.

It's in the dream mill that a brain reconciles the self with the world, an oblique process vital for survival. Other animals dream, too, and we probably inherited the gift from earlier species as a space-saving way to process memories. Flowing through the caverns of the mind, dreams rarely reach consciousness, but sometimes do, especially if one wakes during or just after REM sleep. Then one can tap into a dream, but only using vanishingly brief short-term memory, which is why little of a dreamworld can be recalled.

In part, dreams are the nightly journal of a mammal's memories. During REM sleep the day's trials and the past can be combined to fine-tune behaviors. After long-term studies of the hippocampus, REM sleep, the dream states of other animals, and brain waves, the neuroscientist Jonathan Winson, of Rockefeller University, proposes that dreams reflect "an individual's strategy for survival," both physical and psychological, by "incorporating self-image, fears, insecurities, strengths, grandiose ideas, sexual orientation, desire, jealousy, and love." Studying the echidna (spiny anteater) provided Winson with important clues. Because echidnas don't experience REM sleep, he realized that REM must be very ancient, evolving about 140 million years ago, when our ancestors branched off from the earliest mammals. Winson also found that a special brain wave, theta rhythm, plays a key role in dreams. In animals that are awake, theta rhythm appears as they respond to a changing environment, something that differs with each species, depending on their modus operandi. These aren't genetically ordained behaviors, such as feeding, but responses to novel circumstances, however slight. Theta rhythm appears to be generated in several areas of the hippocampus, all moving in synchrony, with a period of about 200 milliseconds between waves.

Once Winson could trace theta waves from brain stem to hippocampus, a sharper picture of memory emerged. As a rat explores, whiskers twitching, it smells its world, and feedback from all the senses meets in the hippocampus and the region surrounding it, where it's divided into "200-millisecond 'bites' by theta rhythm" and goes into long-term storage. Clearly theta waves play an important role in an animal's waking life, but during REM sleep theta rhythm also conducts the hippocampus and neocortex in their dance. Pity the poor spiny anteater, unable to dream-process the day's information. Without REM sleep, a sort of compactor, our clogged brain wouldn't have spare room for the sensory jumble we enjoy and it relies on to survive.

No one really knows why sleep evolved. But the larger an animal is the less sleep it needs. Ferrets sleep about fifteen hours a day, cats thirteen, humans eight, and elephants only three. Small animals, blessed or burdened by fast metabolisms, would melt down if they didn't snooze away most of their lives. REM sleep may play several roles. Dolphins enjoy only traces of REM sleep, and yet they have good memories. Smart humans don't get more REM sleep, nor stupid humans less. Intriguingly, one thing that does correlate with amount of REM sleep is how helpless offspring are at birth. An opossum needs eighteen hours of REM sleep, and a platypus even more. Does a platypus dream? Is there an opossum's nightmare? For that matter, what does a newborn human dream during its long pastures of sleep? Animals born sprightly and mature can hit the ground running. A newborn gazelle wobbles to its feet and soon trots alertly beside Mom. But immature newborns have many neurons to thatch, vital sensations to unpuzzle. That's urgent, so why experience life only while awake? Maybe the rough and tumble of dreams helps to tune up a baby's senses and prepare it for the teeming carnival ahead, one complete with imperceptible con men and hucksters, and a midway bustling with real and imaginary lies.

CHAPTER 17

"Hello," He Lied

(True and False Memories)

When I was young I could remember anything, whether
it happened or not.

—Mark Twain

While many patterns and interpretations build the *what* and
why of memory, the source of a memory is easily confused.
In a previous chapter's orchard memory, I recognized a pattern as
a plum tree. But how did it get in my brain? Did it get there
because I saw it? Because I heard someone talking about it? Because
I imagined it? Fantasized it? Dreamt it? Read about it? We often
misattribute the source of something familiar.

Source misattribution can be auditory, visual, olfactory. Did
your friend tell you about a UFO sighting, or did you hear it on the
radio? Guilty only of cryptoplagiarism, you may think an idea
original when in fact you read about it a year before. A professor
may honestly believe he invented an opinion that, in reality, his
teaching assistant provided. We sometimes remodel memories.
Digging around, we glue memory fragments together and lose the
source information in what the Cornell psychologist Steven Ceci,
a leader of long-term research in this phenomenon, likes to call the
"tossed salad" of normal memory.

The psychologist Ulric Neisser, also of Cornell, has been testing
this in fascinating ways over the years. In a typical study, begun the
day after the *Challenger* Space Shuttle exploded, he asked people for

their memories of the event. Three years later, he surveyed them again, and about two thirds were totally wrong about where they heard the news, when, with whom, and so on. The people who were dead wrong were just as confident as those who were right. After a northern California earthquake in 1989, he had professors hand out questionnaires in San Francisco, Santa Cruz, and Atlanta, requesting personal memories of the quake. A year and a half later, he surveyed again, and the California students got most of the event right. But the Georgia students didn't. They hadn't been especially upset by the tremor, which they'd heard about in the news, and so memories weren't felt by the body, stamped in through physiological experience. The California students experienced the physical event and also talked a lot about it. Neisser found that if you tell a story about an event, you're more likely to remember it. Narrative, one of the brain's key strategies, helps engrave memory.

Adults frequently make such source errors, but preschoolers do it two to three times more often. Young children and elderly adults are especially vulnerable because the frontal lobe, an active participant in source memory, matures slowly in children and decays quickly as we age. Both groups may be able to remember something keenly, and interpret it, but have amnesia about the where and when. Impressionable, especially at life's twin poles, we're often confused and easily persuaded. So, for example, researchers can suggest that when you were five years old you spent a week in the hospital with pneumonia. Later you attribute the memory to actually being hospitalized, not to the suggestions. Or you might repeatedly be told that you forgot your lines in a school play when you were eight, and later that becomes a stinging memory for you, when the real source is only a suggestion someone made.

This vulnerability lies at the heart of a controversy over memories of abuse. In the case of false memory, a person isn't lying; he or she does believe the memory, even if it was a fiction. Three-year-olds can remember pretty well months later, and their memories can be contaminated. Usurp someone's memory, and he'll grow a new narrative over time, one that thrives in his belief system. What began as a lie gains status as a memorable truth. Tell a lie often enough, and it blends into your autobiography, and when that happens you become devilishly difficult to detect as a liar. Because you're simply recalling what you think happened, you wouldn't feel anxious about lying, your face wouldn't leak clues.

But brain mapping might. Polygraphs only measure the vital signs of someone anxious about being caught in a lie. Brain scans seem to show that when someone is lying, a part of the brain involved in executive functions becomes active. Lawrence Farwell, chief scientist at Brain Fingerprinting Laboratories in Fairfield, Iowa, has been testing a more precise lie detector, a controversial technique that maps P300 brain waves, which the brain produces as it recognizes the familiar. When Farwell scanned the brain of a convicted murderer appealing for a new trial, he found that details of a particular crime scene didn't stimulate the same area of the man's brain—that is, didn't register as familiar, while details of his alibi did. Although the results suggested his innocence, the appeal wasn't granted. Not remembering a crime doesn't prove innocence. The Lockheed Martin scientist John Norseen, financed by the Department of Defense, is trying to map the brain waves of guilty thoughts; he predicts that airports will one day scan everyone's brain and detain terrorists . . . or a man feeling guilty about lying to his mother.

It turns out that you can do one thing to really clobber source memory—add stress. Then memory falters, and the source part suffers. So let's say, for the sake of argument, we want to create source errors on purpose. We could use any or all of the time-proven ways known to scientists and dictators alike: create expectancies, use leading questions, criticize or punish wrong answers, reinforce the answers we want. By suggestively questioning someone, we could instill expectancies. This is the "When did you stop beating your wife?" sort of question. We could also reward someone—materially or emotionally—for answers we prefer. We could ask someone to use only special imagery, which would limit his or her responses. We could impose stress. Or we could combine all of the above.

Like light traveling a great distance through the cosmos, memory bends. Encourage someone to picture something long enough, and she'll confuse what she's imagining with what really happened. Children can hear persistently, for example, that a person, culture, or race has bad qualities, and believe they saw them firsthand. Or someone could ignore the children entirely and focus on their parents, advisers, or interviewers. Searching for proof of their own beliefs, adults might unwittingly sculpt a child's memories. "It's not that kids are hypersuggestible sponges," Ceci cautions. "You have to work hard at it to get false memories." It takes repeatedly suggesting something over months and years, but that's precisely what happens in homes, neighborhoods, and schools.

In addition to false memory, there is the curiosity of false forgetting. One morning an anesthesiologist from the Hospital for Special Surgery in New York City phoned to inquire if I was having any postoperative problems from the ankle block and anesthetic I'd received the week before during a foot operation. I wasn't.

"Just out of curiosity," I asked, "what drug was I given to knock me out? I went into a deep sleep, but I woke up alert when the operation was over."

"Oh, you weren't knocked out," he said. "You were given versed—it erases your short-term memory."

"I wasn't asleep?"

"No. You just don't remember the operation."

"Wait a minute. I was awake and talking, but I don't remember anything I said?"

"That's right."

"Doesn't this sound like a good scenario for blackmail?"

He laughed. There was no point in my asking him if I said anything outlandish or embarrassing; he'd deny it. One fears the worst. Latent Tourette's syndrome. Tedious meanderings about one's lumbar region. What happens to the superego when one is in that alert but forgetting state?

"I'll tell you one problem, though," he said. "A patient will ask the surgeon a question, receive an answer, but then, because the patient can't remember, ask the question over and over again . . . until finally the doctor tells me to increase the dose."

I can understand why doctors might wish to use a memory

block. Some patients, dawning from full anesthesia, have reported hearing surgeons' talk. Humor can be crude in an operating theater. In one operation I observed up close, a male surgeon missed the eye of a needle he was trying to thread and wisecracked: "If it had hair around it, I wouldn't have missed." But I suppose it's worse if the patient hears a surgeon saying something like, "Not much hope for recovery with this one, is there? I'd hate to be in her shoes," or even simply, "Oops!" So, blitz to the memory. Temporarily, at least.

Not knowing you've forgotten isn't as bad as knowing you've forgotten. Then there are the silent persuaders we meet in advertising campaigns. Repeating something on purpose helps one to remember it. But unconscious repetition works, too; indeed, it works alarmingly well to influence our feelings, hopes, and desires. Subliminally, the brain registers the information. Show a film of someone while negative words flash too fast to notice, and a moviegoer may report feeling hostile about that person. In one experiment, interviewers talking with elderly subjects tended to walk more slowly, with curved shoulders, when they left the room. Invisibility can take several forms. The first might be fully present but abstract, like an agreement or math problem. But there's another sort of invisibility like the sky shapes carved by buildings, or a friend out of sight, or the wallpaper in a dining room, which we register but don't really notice. A subliminal invisible indivisible from thought. Advertisers lard our days with yearnings for their products but, more insidious by far, also with values and prejudices.

We dig through the sands of experience and assemble memory fragments into a plausible whole, sometimes losing their bona fides. Some of them, mistakenly, we claim as our own. Did I read that or think that? Did that happen to me or someone else? Phantoms stalk our memory. We net some in sentences, while others shimmer away. Our biggest fairy tale may be the truthfulness of so-called reality, when hidden fictions steer us through the darkness of our days.

CHAPTER 18

Traumatic Memories

Time is the substance from which I am made. Time is a
river which carries me along, but I am the river; it is a
tiger that devours me, but I am the tiger; it is a fire that
consumes me, but I am the fire.

—Jorge Luis Borges

The gold of memory may be panned from experience, but it's
washed in the acids of emotion, which etch the deepest memories. For years, whenever I heard Vaughan Williams's gorgeous
"Fantasia on Greensleeves" I felt frightened and anxious because
that music happened to be seeping, almost inaudibly, from the ship's
radio in the days following an accident at sea in the South Pacific,
a horrifying event where people I knew were thrown onto a coral
reef and some injured, some killed, while several of us tried to save
them in the jolting surf. Like a film, the memory of that event has
run across my mind, unwanted and unbidden, countless times
since then. It happens as if for the first time, in panoramic sound,
smell, taste, feelings. Part of the trauma, I think, had to do with how
many sensations were flagging the memory and its steep emotions.
I can still hear the deafening bronchial sound of the surf, with the
outline of screaming voices superimposed on it like transparent
sheets in a *Gray's Anatomy* of fear. I can still feel the sharp teeth of
the man to whom I gave mouth-to-mouth while another com-
pressed his chest. His teeth cut into my gums, and seawater frothed
up from his lungs, as I breathed for him for what seemed ages, until
I knew I'd failed him for good, forever. He'd been thrown ashore by

the surf, and probably was already dead when we began trying to revive him, but surely there was something I could have done? I can still see an elderly man who staggered drunkenly out of the surf with all of his clothes ripped off him. As I recall, his wife was airlifted out. Did she die? I used to know. Finally, after a decade, the event doesn't rattle the present with its visceral sensations but is stored as a vivid memory. I partly accomplished that myself in a curious way I'm glad to share, in case it may work for others.

One day I decided to take a Red Cross CPR class, since I've a loved one with a heart condition. I dreaded the class, but dreaded a coronary emergency more. After only a few minutes in the class, flashbacks gunned and I was panicky, my heart galloped, my nerves felt like they were all flinching in unison, and I wasn't sure I'd be able to stick it out. Wanting to run from the room, I remembered why I was there and forced myself to stay. A loud flashback raged inside my eyelids. Crash and caterwaul of the waves with human screams ricocheted around the room, though no one else seemed to hear them. Could people tell? Was I shaking? *Don't run,* I ordered myself. *See if you can turn off one of the senses. Try sound.* I concentrated hard on turning off all the sound, only sound. It worked. The flashback became a silent film. Doubting I could practice mouth-to-mouth on the dummy, I concentrated on turning off touch—not the feel of the dummy, only touch in the flashback. The sharp cutting teeth. The man's sudsy death hitting my mouth. The bizarre sexuality of something like open-mouthed kissing with a man, but in the service of death. *Turn off touch,* I demanded over and over, until it numbed. For me at least, having even partial control of a traumatic memory took some of its ferocity away. That, combined with psychotherapy, worked wonders. When I think of the event today, it's as memory. Something may remind me of it, and I may feel chilled or sad, but it's not in the now so much, not the terrifying emotional red alert it had been.

When any danger looms, memory does two things immediately. It quickly records where you are and why, who you are with, what you feel, the full pandemonium of wretched details. Those are stored consciously by the hippocampus. At the same time, on a more subterranean level, the amygdala records sights, sounds, and other perceptions, and it files those as new signs of danger. From then on, every time you encounter those stimuli the amygdala surges and you feel fear.

On the molecular level, here's what happens: a stimulus associated with a frightening event—let's say hearing "Greensleeves" on the radio—releases a neurotransmitter in the amygdala, which in turn draws calcium into the cell, which prompts special proteins to go to the nucleus and turn on the genes that make other proteins, which then return to the synapse where the music was being processed and glue that music to the site. So whenever the music plays, the synapse comes alive and the cell fires. Or, if you like, a memory happens. The more the memory is rehearsed, the stronger it becomes. I don't mean to villainize our friend the amygdala. Part of the amygdala's job is to override instincts if need be and remind the brain of how the world *feels*, good or bad. The succulence of biting into a sweet ripe apricot. The water-swallowing panic of nearly drowning.

Our genetic memory equips us well with *how* to feel fear—limbs freeze, the pulse races, blood pressure soars. Mainly through experience, the brain learns *what* to fear. And, of course, there are many false alarms, many harmless things that trigger fear through what's essentially guilt by association. It only takes a familiar scent, shadow, or phrase. If a man was cooking pancakes when he learned that his brother died, making pancakes might always tinge him with sadness, though he may not know why. The brain expects to find a trauma where it happened before, and with some of the original paraphernalia. The crisper the details, the more gut-wrenching the memory. Essentially the trauma is being anchored with sensory ropes, by powerful memory aids. With help, a past threat can be stored where it belongs, in memory's attic. But razory sensations tend to stay sharp and undecayed, unless they're consciously smoothed by repeated handling.

James McGaugh, of the University of California at Irvine, discovered that if he gave rats a shot of adrenaline after they learned something, they remembered it better. Adrenaline beforehand didn't work. Only afterward did it boost memory. McGaugh suggests trying this with trauma victims, giving soldiers and rescue workers beta-blockers like Inderol immediately after they've experienced horror to soften their memory of the event. They'll still remember it, but without the emotional fireworks. Interestingly, when adrenaline soars during a frightening event, it heightens romance. After a scare, men and women describe each other as more appealing, and they're more likely to ask for a date.

Most children don't remember much before the age of two or three, because the cortex develops slowly and the brain hasn't activated its thinking cap yet. But they can store strong emotions unconsciously. The brain remembers early emotions and sensations as flighty feelings and images rarely netted by words. It takes about eighteen months for the language systems to mature, and even then a child understands words before she can speak them.

Early fear and abuse scar in subtle ways. We learn in childhood how to decode faces. Over time, abuse can program the senses to detect even a cloud of anger on someone's face. In a study of physically abused and nonabused nine-year-olds, the abused children read anger in expressions that weren't predominantly angry, and they had trouble telling angry faces apart. Expecting anger, and finding it, even when the clues are paltry, could poison school for such children, ruin friendships, undermine trust. In a similar way, children of depressed parents are able to read tinges of sadness on faces, sometimes even imagining it, so habitual is their search for clues.

A clever escape the brain devised was flight in place: imagining alternate worlds free of terror, then completely melting into those worlds or making spa-like forays into them for relief. Imaginary worlds may contain their own goblins, but they can be bridled, scripted, carefully dosed, and ultimately resolved. Adding a bonus to escape fantasies, the brain can sense imaginary events as real (hence the success of athletes visualizing perfect performances). When I was little, I followed that Silk Road of the imagination. Although I couldn't name the terror or understand its origins, I feared igniting my father's rage, feared my inability to quench it, feared my mother's desertion. Instinctively, my brain threw me a passport to other worlds. Small daily acts of terror happened (and were recalled) as raw sensation, while my mind kited away to another now/here land. I slipped between the real world and my secret one often, undetected, sometimes midsentence, as if behind sliding paper doors in a Japanese guest house. It's a good thing the imagination, which responds so well to bribery and invocation, can also appear unbidden when needed. That compound ghost, haunting our synapses, sleepily at times, and at others with fierce resolve, stays limber by amusing itself in endless antics and heavy lifting, muscles ready for the next steep jagged descent.

Of course, imagination doesn't only function as survival's valet. Sometimes the servant becomes saboteur. It can take a terrifying instance and turn it into an everywhere. I'm not sure why trauma lures, but there's no doubt the mind likes to fix on a real or imaginary horror and replay it in gruesome detail. There are atrocities I would never imagine, but, learning of them, can't forget. They sear memory and leave a scab my mind returns to and picks at. I dare not read about or view atrocities, because they'll become indelible. They'll lodge in my neural circuits, leap out at random moments and terrorize me. Years ago, I learned of a World War I atrocity—a German soldier bayoneting a baby out of a pregnant woman's stomach—that still flashes into mind and leaves me scorched. Even typing that sentence—a bare-bones report of a complexly gruesome act—felt painful. Emotional quicksand accompanied it, and I sank briefly into the horror that woman must have endured.

To some extent, the brain seems to provide its own traumatic memory softeners. Survival depends on regaining stability, readjusting, forgetting, welcoming the new, and relearning. Beat Lutz and his colleagues at the Max Planck Institute of Psychiatry in Munich have found that mice bred without a certain brain system (cannabinoid) respond differently to trauma than normal mice. If normal mice are given a shock when they hear a bell, they soon learn to fear the bell, anticipating the shock; when the shock is removed, they learn the bell is harmless. But Lutz's fright mice stay scared long after the shocks stop. They can't seem to forget the trauma, even when the environment changes and safety reigns. "This could explain why some soldiers recover, while others end up dealing with shell shock for decades," Lutz offers. How prone one is to fear, and how long one stays afraid, may be genetic. In NIH studies, people shown fearful faces felt more anxious, and had more activity in their amygdalas, if they happened to possess a certain variant of a gene controlling serotonin.

Should we try to erase all memories of a trauma? When traumas *don't* get stored as memory they can stay alive in a brutal present of pure sensation. But as memory, they can enrich one's identity, since memory provides a renewable sense of self. I may partly define myself as someone who didn't hesitate to try to save drowning people, and someone vulnerable enough to be traumatized by their deaths.

CHAPTER 19

Smell, Memory, and the Erotic

Smells are surer than sights and sounds to make your
heart-strings crack.

—Rudyard Kipling

I begin each summer day in earthly paradise, tending my roses,
which take over an hour to preen, because there are 120 bushes
in several beds, and I raise them organically. I admit, that's a lot of
roses, but I feel that it's only obsessive if I count them each day. I
prefer to think of it as a floral exuberance. Even though florists' roses
rarely smell (they're bred for color and shape, sacrificing scent in the
process), many of my garden roses ooze a perfume heavy as gold
ingots, elaborate, intoxicating, and cunning. What pollinating bee
or human could resist them? While tending, I admire them like an
acolyte, listen for wren hatchlings, watch the sun seeping through
the woods, enjoy the rustle of tree leaves, and try not to let anything
as clumsy as thought intrude. After that, the other flower beds call.
I find these edge-of-the-morning hours restorative as meditation.
When I recite my garden as a living mantra, I'm drenched in aromatic memories, especially the sense-luscious smell of roses called
Othello and Abraham Darby.

I'm fascinated by the way we humans like to bathe ourselves in
herbal or floral memories: lemon zest, rose petals, gardenias, honey,
sprigs of lavender. We want to luxuriate in sensual memories, and
also tell anyone we encounter—even strangers, fleetingly—that
we're as tantalizing as blooming nature. We want to conjure up a
season of ripeness, the sexiness of flowers sticky with nectar and

succulent fruit. My favorite perfume at the moment, Obsession, relies heavily on vanilla for its allure. Despite its adult name and sexy film noir advertising campaigns—which create their own subliminal appeal—its scent returns me to the kitchens of childhood, since vanilla figures heavily in baking. Another scent in the category of pure nostalgia is talcum powder, important in many perfumes because baby boomers' parents talced our bottoms. Our brains blended those first smells with emotions long ago, and, as advertisers have learned, a smell can retrieve hidden memories and speak to the unconscious. Used-car dealers spray an eau de new car scent in their reruns, and restaurants have been known to pump aromas into shopping mall ventilation systems.

When we give flowers, scented soap, or perfume to someone, we're really giving them packaged memory. Unlike our other senses, smell is linked very closely to memory, and easily washed with emotion. Most sensory details go first to the thalamus, the gateway to the forebrain. During sleep, the gate closes or we'd be too fretted by sensations to rest. But when the nose detects something, it bypasses that route and sends a message straight to the limbic system, a mysterious, intensely emotional part of the brain, full appetites and urges. Both the amygdala (emotion) and the hippocampus (memory) take note. Few things are as memorable as a smell, which can be overwhelmingly nostalgic because it triggers powerful images and emotions before we have time to edit them. What you see and hear may quickly fade into the compost heap of short-term memory, but there is almost no short-term memory with smell. We each have our own special aromatic memories, in part because smell stimulates learning. In experiments where children are offered scent cues along with a word list, they remember the words better.

We're lucky to live on such a fragrant planet. Some of the richest scents don't ooze from flowers, but from leaves and bark, and even the atmosphere itself. I love the smell of spring in the air at the end of winter, a delicate, fertile world-in-bud smell. Many spring bulbs offer scent, from the nose-bludgeoning hyacinth to the faintly lemon poet's narcissus. All winter I wait for the magnolia and wisteria to bloom. The flowers of the wisteria smell slightly Easter lily-ish with a hint of new car leather, and I like to stand with my nose against a cascade of wisteria, inhaling its heady perfume for as long as possible before my nose quits and refuses to smell it anymore. My technique with the magnolia is similar. I have a large old Chinese magnolia with brandy-snifter-shaped pink flowers. Although it blooms only for a week or two, it's an otherworldly spectacle. At some point each day I walk straight into those flowers, pressing my face into a thick mask of petals, and then breathe slowly and deeply, enjoying their gentle scent of vanilla and warm candle wax.

Something about the rich changing smells of a garden comforts and exhilarates me, and also fills my unconscious with seasonal memories. Most summer mornings I indulge in a scent tour of the garden, pausing to inhale the nearly anesthetic scent of some lilies and hostas. Smell something too long and habituation sets in, that *Ho-hum, another sword swallower* message the brain uses when nothing's new and it might as well idle. At that point, the nose seems to quit, the way a muscle tires from overuse. But after a little rest, the nose is back in business, ready for another plunge of lilac, Thai basil, or cedar. Then it's time to stroll about the garden for a while longer, inhaling with a rested nose.

A small grove of white lily of the valley tolls miniature bells of scent, a perfume somewhere between musk and apples. I received the ancestors of this crop from my parents decades ago; they were transplanted from the backyard of the house I grew up in, and their smell trails memories of high school years. I also grow herbs. Although I don't bottle the smells, I use them as perfumes, just as women have throughout the ages. For example, I've planted a peppermint patch in such a way that I have to walk through it when I want to turn on the garden hose. Then a cloud of peppermint— as much vapor as scent—washes over me, almost always catching me by surprise, reminding me of adolescence once again. Putting peppermint leaves into a tea ball and tossing it into the bath creates a nose-tingling soak. I like to stuff lavender, rosemary, sage, or lemon verbena into sachets for sweater and sock drawers. In these ways, I greet nature each day with a sensory treat before my work begins. Then, when winter turns the town into a toboggan run, and the radio warns: "Don't go out, it's dangerously cold!" I can simply open up a drawer, pull out a sweater, and clothe myself in the perfumed memory of last summer.

This wouldn't surprise Rachel S. Herz, who studies smell and memory at the Monell Chemical Senses Center in Philadelphia. In an early experiment, she had subjects inhale certain odors while looking at paintings. Several days later, she either gave them the smell or mentioned the word for it. They responded equally well to both, remembering the painting on cue. But only the smell produced an emotional memory that allowed the subjects to remember what they were feeling when they first saw the painting. Smell also increased their heart rate. Since then, Herz has conducted many other experiments pointing to the emotional power of smell. "I really believe that olfaction and emotion are the same thing on an evolutionary basis," she says. "I think emotions are just a kind of abstracted version of what olfaction tells an organism on a primitive level. And that is why I think odor has such a potent emotional cascade."

Like most people, I have scent memories enough to fill an entire novel, but I wouldn't be the first person to use smell to delve into the past. We tend to remember a past rosier than real. Some details loom and others shrink; we simplify, exaggerate, add to, rationalize the memory. Memory can make a moment larger and more important than when it actually happened. At first a bad memory may hurt, but then slowly grow, fill with associations, provide nourishment. No one knew this better, or told it more sensually, than Marcel Proust, the great caresser of memories, the voluptuary of smell. In a short story set in a depressing grand hotel, he writes of the scented secret of Room 47, obviously inhabited by gods of love temporarily calling one another Violet and Clarence. Walking down a corridor, he's surprised by "a rare and delectable scent." He couldn't identify it but found it "so richly and so complexly floral that someone must have denuded whole fields, Florentine fields, I assumed, merely to produce a few drops of that fragrance. The sensual bliss was so powerful that I lingered there for a very long time without moving on." The door was open barely a crack, just enough to allow the stirring fragrance to escape, not enough to see or analyze the personality of the guests. Still, he wondered how, staying at that nauseating hotel, they managed to sanctify their boudoir, turning it into a fragrant oasis. Later he encountered the perfume once more and followed its rippling back to the now disheveled empty room of the recently departed lovers, so blended at the level of cell and molecule and yet so jealous of their intimacy that they couldn't bear the idea of anyone sharing their private language of scent:

> I was numbed by the violence of fragrances, which boomed like organs, growing measurably more intense by the minute. Through the wide-open door the unfurnished room looked virtually disemboweled. Some twenty small, broken phials lay on the parquet floor, which was soiled by wet stains. "They left this morning," said the domestic, who was wiping the floor, "and they smashed the flagons so that nobody could use their perfumes."

Although they couldn't fit the flagons into their crammed trunks, the perfume had lavishly scented their memory, and they didn't want anyone eavesdropping on their passion. But Proust's narrator finds a few drops left in a vial and takes it home to scent his own room. And what memories does the perfume conjure up for him? Loss and consolation:

> In my humdrum life, I was exalted one day by perfumes exhaled by a world that had been so bland. They were the troubling heralds of love. Suddenly love itself had come, with its roses and its flutes, sculpting, papering, closing, perfuming everything around it.... But what did I know about love itself? Did I, in any way, clarify its mystery, and did I know anything about it other than the fragrance of its sadness and the smell of its fragrances? Then, love went away, and the perfumes, from shattered flagons, were exhaled with a purer intensity. The scent of a weakened drop still impregnates my life.

Proust was born in Paris in 1871, at the climax of the Franco-Prussian War, a time of hideous deprivation, short rations, and disease. Unable to get the nutrition she needed during pregnancy, his mother blamed herself for Marcel's frail start in life. He stayed bedridden for much of his childhood, often missing school, and his mother nursed him while his doctor father worked. These were golden days of love and discovery for young Marcel, whom his mother teasingly called "my little wolf" because he devoured her care, a time when the sun always stood at noon, and he monopolized the love of the only perfect creature on earth.

As an adult, Proust moved among the highest echelons of Paris society, but he spent most of his life under covers in the cork-lined bedroom of his sumptuous apartment. Chic, witty, wealthy, cheerful, a dandy, full of gossip, obsequious in the extreme, he was nonetheless frail and ill (he died of asthma at fifty-three), but he was also emotionally in retreat. The Midnight Sun, his Parisian friends called him because his hours were reversed: he slept by day and wrote or socialized by night. Nearly a hermit, he lived in a night land remote as deep space. It was there, in his palatial rut, propped up against exquisite pillows, eating mashed potatoes delivered from a favorite posh restaurant, that he created his masterpiece of embellished recollection, *Remembrance of Things Past,* in which he tried to remember everyone he had known, every self he had been, everything he had seen or done for his entire life. How can one convey the ampleness of being alive—all the people and emotions, animals, skies, sensations and thoughts, as well as the subterranean life of the mind itself? His fictional frieze sprawls for three thousand pages, whole sections of which sing with the gorgeous music of the mind and heart and are, appropriately, unforgettable. "He was a dream analyst," the novelist Paul West writes, "a trance-conjurer, a scandal-savorer, a snob maven, a dealer in smart remarks, a prodigious theorist of love's memory."

The adult Proust didn't search for childhood memories to mine. They came unbidden as manna, and he referred to them as involuntary. That is, they weren't drafted for novelistic service, they just happened. But once they did appear, he turned each into a small forever, a mini-universe of inexhaustible study, a carousel of sensations. In *Swann's Way*, to use the famous example, on a cold winter day, Marcel's mother offers him some scallop-shaped "petite madeleine" cookies and tea. He soaks a morsel of cookie in a spoonful of tea and raises it to his lips. When he tastes it a shudder runs through him, a gong sounds in his memory, and he is transported to his childhood visits with his aunt, who served him petite madeleines and lime-blossom tea. He retastes those plump cookies, resmells those cups of fragrant tea. A dam has opened and a river of textures, atmospheres, sights, and sounds flows in. Blessed with a nearly photographic memory and a passion for accurate detail, he is able to paint his sensations onto the reader's mind so powerfully that each reader feels slid into the room with Proust's aunt and her maid, an intimate part of the scene, all alone, as if no one else on Earth had ever read or imagined it. A voluptuous animist, Proust believed that memories hid like demons or sprites inside objects. One day you taste a cookie—or pass a tree, or see a bow tie—and the memory leaps out at you. When it does, it unlocks the door to all the memories surrounding it, and a sensory free-for-all ensues. The past is a lost city of gold—complete with fabulous temples, quixotic rulers, mazy streets, and sacrifices—that can be discovered in all its grandeur.

Only the inaccessible and elusive is truly alluring, Proust says. And what could be more inaccessible and elusive than the past? Each person is attracted over and over again to a predictable "type" of lover. Each has a habitual pattern of loving and of losing: "the men who have been left by a number of women have been left almost always in the same manner because of their character and of certain always identical reactions which can be calculated: each man has his own way of being betrayed." For Proust, love is a conscious, deeply creative act of communion with memory, reaching into and through the beloved to all of life. As he says, "The fact is that the person counts for little or nothing; what is almost everything is the series of emotions, of agonies which similar mishaps have made us feel in the past in connexion with her." We do not love people for themselves, or objectively; quite the contrary, "we alter them incessantly to suit our desires and fears . . . they are only a vast and vague place in which our affections take root. . . . It is the tragedy of other people that they are to us merely showcases for the very perishable collections of our own mind." Accordingly, it is only because we need people in order to feel love that we fall in love.

Lamenting the loss of his mother, his sweetheart Albertine (the first person to die in a plane crash), and others, he understandably wondered if time were irrevocably lost. But he replayed love voluptuously in his mind, caressed the memories with his pen. So despite Proust's pessimism, he contributed deeply to our understanding of how loving memories feel. He traced the patterns of relationships, and showed how each fresh heartache resonates with past ones, making our "suffering contemporaneous with all the epochs in our life in which we have suffered." Once the beloved is gone, through death or abandonment, grief fills all the seams of one's life. But ultimately, if we wait long enough, grief will become oblivion. For Proust, each stage of love bridges time and is colored by a sensuality all its own, especially the final stage—waiting through grief for oblivion—which is perhaps the most welcome of all, since it restores one's sanity until the next emotional uprising. As Virgil wrote in the *Eclogues*, "Time bears away all things, even the heart." Meanwhile, scented memories veil us in rapture.

NEVER A DULL TORMENT

(The Self, and Other Fictions)

CHAPTER 20

Introducing the Self

If "I" give my love to you, what exactly am I giving and
who is the "I" making the offering, and who, by the way,
are *you?*

—Stephen Mitchell, *Can Love Last?*

A self is deciduous, it leafs out as one grows, changes with
one's seasons, yet somehow stays briskly the same. The brain
composes a self-portrait from a confetti of facts and sensations, and
as pieces are added or removed the likeness changes, though the
sense of unity remains, thanks to well-furnished illusions. We
need illusion to feel true. A medley of different selves accompanies
us everywhere. Some are lovable, some weird, some disapproving of
each other, some childish or adult. Unless the selves drift too far
apart, that solo ensemble works fine and copes well with novel
events. As the psychoanalyst Philip M. Bromberg writes in *Standing in the Spaces:* "health is not integration. Health is the ability to
stand in the spaces between realities without losing any of them.
This is what I believe self-acceptance means and what creativity is
really all about—the capacity to feel like one self while being
many." Even at the cellular level we're a mosaic. A self is a powerful sleight of mind arising from 100 billion neurons communing at
100 trillion synaptic bridges.

Without a self, the complex socializing we require to finish
wiring our brain and teach it survival skills and mating wisdom
would be overwhelming. Just think about the complexity of
speaking with another person, someone important to you, in a

relationship you cherish. You're attending to the subject you're talking about, talking, and, at the same time, monitoring how the relationship is faring from moment to moment.

As Virginia Woolf writes in her novel *Orlando,* where a character's selves occur like sheaves of grain, "A biography is considered complete if it merely accounts for six or seven selves, whereas a person may have as many as a thousand." A long, ghostly parade of previous selves trails behind us, as values, habits, and memories evolve to better reflect the current *I.* We often translate how that feels into spatial terms, and refer to our different facets, or sides. All of our selves seem to inhabit separate spaces. The mind needs spaces to juggle its different concerns at once, which sometimes are in sync, sometimes not. When they're not in sync, there has to be a way to proceed fluently, without stumbling every time there's a rift in what one part of you is conscious of emotionally and the other part is conscious of cognitively. The brain needs to separate the experiences enough so that what's most important at each moment can be preserved. If an inner kid is being disruptive, we might move into one mental space to soothe it as quickly as possible, but quite another to figure out what's best for it. Most of the time, those spaces aren't terribly significant, unless one had early experiences that made keeping them separate indispensable for survival. Then one might spend the rest of one's life making sure they always remain separate. A self is the trail of bread crumbs we leave so we'll know our progress and direction. William James envisions the self as "the sum total of all that [a man] can call his, not only his psychic powers, but his clothes and his house, his wife and children, his ancestors and friends, his reputation and works, his lands and horses, and yacht and bank account. . . . If they wax and prosper, he feels triumphant; if they dwindle and die away, he feels cast down."

Wide and rich as a self is, a feast of being, surely it deserves more than a mere four letters? The word *self* meanders back to the Indo-European *s(w)e-,* which gave birth to countless offspring, such as *gossip, suicide, idiot, desolate, sullen, sober,* and *swami* (Sanskrit for "one's own master"). Because a self is a mirage made from so many sighs, it's hard to pinpoint its whereabouts in the brain. But scientists have done a good job of locating its general neighborhood. Psychosurgery of a brutal sort led the way.

When frontal lobotomies were fashionable (in the 1940s, about twenty thousand were performed in the United States), doctors and families alike felt relieved by how obedient and sociable their depressed, manic, or schizophrenic loved ones became. Rough-and-ready traveling surgeons specialized in knocking off several lobotomies a day. Here's one surgeon's account, borrowed from Edward Shorter's *History of Psychiatry:*

> [There's] nothing to it. I take a sort of medical icepick . . . bop it through the bones just above the eyeball, push it up into the brain, swiggle it around, cut the brain fibers and that's it. The patient doesn't feel a thing.

Chillingly true. Afterward, patients often felt oddly bereft of emotion, and some became aggressive (at least one lobotomy patient shot his doctor). Patients suffered less on the whole, but the surgery butchered their sense of identity.

The conscious, preconscious, and unconscious conspire to create the notion of self. A self can be mischievous, a union of impulses and moods. It shuffles constantly, kaleidoscopically, merging the sensations of a moment, and then mixing again, so that it's never quite the same. Using a PET scan to record a typical moment in the brain, we'd find surprisingly few thoughts busied with sights and sounds, smells, tastes and touches. Most of its hubbub originates inside, in mind theaters, fantasies, mental scratch pads, inner monologues, memories, emotions, the baroque architecture of self. From moment to moment, the self states mutate. We're usually unaware of a different self taking the helm, but that helps explain the phenomenon of changing one's mind. One mental state can make a decision disavowed by a later state. Changes occur frame by frame, and, speeded up, they seem to flow like a single film.

One may be the loneliest number, the mainstay of individuality, almost synonymous with self. But one never exists in the brain. As solitary as we feel at times, alone and unknowable in the fullness of our desires, every "I" is a "we," more clan than family, an ensemble of cells. A self is plural. To think of myself as singular, a choir of neurons living in different hills and hollows of the brain must sing in concert. Word travels fast in the back hills of the mind, through the hippocampus, that campus where memories are schooled, and the hobgoblin amygdala, spoiling for trouble, until a chorus of neurons becomes a thought. Sometimes several choruses.

Much of a self derives from recollected events, their weight and outcome, and the personal iconography they create. Since others figure in those mementos, and the daily acts that impress new memories, other people become integral elements of oneself, an important part of our inner diary and identity. If a loved one dies, one loses portions of self, not just *a portion*, because the missing person also hosted different selves. One loses the parts of self linked to the different parts of the loved one; they seem to have fled and can take a long while to restore. When that restoration is hard, it can interfere with mourning, as if portions of oneself really did tumble into the grave. Mourners often imagine the loved one materializing on high, from the back of beyond, looking down as a saintly guardian. Sometimes one even seems to become the lost person, playing both roles. Grieving birds can do that, too. Canada geese mate for life, with the male singing one part of their song, the female the other. After a mate dies, the survivor sings both parts to keep the whole song alive. If humans did that, we would call it romantic.

One summer morning, missing a friend away on vacation, I wrote him a poem. Set in my garden, the poem is a roll call of a dozen or so selves that flourish when he's around. While writing it, I was consciously aware of garden and emotional data, including that the poem was addressed to a special reader whom I wouldn't see for nearly two months. I imagined his self, in some of its complexity, hovering as I composed. Apropos of the relationship, I turned my gaze inward to reflect on several of my personas, choosing ones I thought might be muted by his absence. All that was conscious. But the preconscious had a field day, too. No doubt it was calculating how each phrase and line might affect him and enrich or threaten our relationship. I hoped it would enrich it, but art is always a gamble in which the stakes are often higher than the pay-off. While my conscious and preconscious mind whirred, my unconscious busied itself with who knows what. Fear that our intimacy couldn't stand such a separation? Jealousy that he'd be spending so much time with others? Resentment that he had other loved ones? Thoughts about my father, who had died the year before? All of that three-tiered operatic bazaar, existing simulta-neously in a mind generated by my brain, colluded to form the sense of coherent self I knew and loved and could take for granted when I arose each day. A self isn't something one has to re-create after lunch. Life would be far too exhausting. That's why I call it a *sleight of mind*. It's an illusion that's both efficient and worrisome, sometimes only a hairsbreadth more efficient than worrisome. For most people it works, greases the machinery of living a reasonably long life, full of gobbling and squirting, until, quite suddenly, it's not needed any longer, one drops dead, and the next batch of illusion-ists carry on.

Cynical, does that sound a touch cynical? In middle age, my planning, predicting brain can see that the light at the end of the tunnel is an oncoming train, that the writing on the wall may be a forgery, and its conscious-preconscious-unconscious self is trying to devise a plan for dealing with something it's never encountered before: annihilation. "God made everything out of nothing," Paul Valéry wrote in *Oeuvres II*. "But the nothing shows through."

What else creates a self? That the brain doesn't replace all its cells from time to time, the way a shark replaces its teeth, allows a self to ripen. Otherwise, what would become of the mental souvenirs that help define us? Not only keepsake experiences, but individual features, habits, and values would perish. Minting new brain cells works only up to a point. Add salt water to fresh and you change the ecology of a marsh. Add enough new stockholders to a board, and you change the will of the company. An agile brain mainly grows new connections, not a flourish of new cells. It changes while staying the same. Which is not to say that designer brains won't one day grow neural patches. How much brain gardening could one risk to heal an ailing brain without losing the self? A delicate balance would be needed, but legal and ethical dilemmas would bloom anyway.* Legally, how do we define a person, by body or brain? Suppose we change the face—is it the same person? Suppose we change the gender? Is a transsexual the same person before and after the operation? Is a child the same person as the adult it becomes? How about an amnesiac? Should one be held responsible for acts committed by previous selves? If you commit a crime as a child, and are brought to trial as an adult, should you be tried by juvenile or adult standards? The real wonder of the brain may be the ease with which it crafts a fluent, persuasive, stable sense of self.

*The first conference on "neuroethics" met in San Francisco in May 2002, bringing together over 150 neuroscientists, bioethicists, doctors of psychiatry and psychology, lawyers, public-policy makers, and philosophers to discuss the fascinating legal, ethical, and social implications of brain research.

CHAPTER 21

The Other Self

If you think your body and mind are two, that is wrong;
if you think that they are one, that is also wrong. Our
body and mind are both two *and* one.
    —Shunryu Suzuki, *Zen Mind, Beginner's Mind*

Our sense of self doesn't haunt the brain alone. It's really a compound ghost created in cahoots with the immune system, the body's other storehouse of memories. Most of us feel intact and continuous, despite the changing weather systems in our cells. When a friend greets us with "How are you?" we don't answer with a current report from the organs, tissues, and synaptic junctions. But that would be more accurate. The thinnest membrane seals us in, giving shape to roaming pastures of flesh, bobbing organs, cellular factories, neurons fizzy with electric. Our porous skin bends and breathes; through it the body converses with the world. But the moment we nick that fragile sack of chemicals, invaders rush in through the cracks, eager to render us down to our constituent molecules and feast. Carnivores can be small, yet ruthless and implacable. Most of the world's carnivores aren't even visible to the naked eye, and yet they make up the largest biomass on Earth, especially in ocean water, but throughout our watery coves, too. Some bacteria are helpful symbionts; some viruses seem to have influenced our evolution. Complete ecosystems on the move, we house four hundred types of microbe in the mouth alone, along with mites nestled among our eyelashes, sweeping hordes of bacteria (10^{14}+ of them), fungi, protozoa, and other hitchhikers. The number soars when we share our fauna by cuddling and kissing. Most of them don't bother us. Others pollute and putrefy as they devour. Only the thinnest fabric stands between us and desolation. Because the body needs to target enemies on sight, the immune system assembles a rogues' gallery of villains, and compares portrait with invader before it launches an attack. Because it's tipped off by resemblances, it sometimes mistakenly shoots itself in the foot, blasting its own tissue.

Long before any threat appears, the immune system erects a wall between self and other, safe and not safe, inner kingdom and outer. Because we must allow entry into the castle of our body, the immune system patrols the borders, and also guards internally throughout the organs. The brain knows who we are. The immune system knows who we're not. Together, trillions of cells build a mosaic memory of the self they defend, a watery being that metamorphoses with age, stress, and mood. A troubled mind can trigger hormones that may subdue the immune system at the worst possible time (when a loved one has died and we're already at low ebb). A mistake can muster a self-attack.

As I write this, I'm vexed by a thin, smile-shaped ridge of inflammation around my mouth, a condition unaffectionately known as *erythera areafa migrains*. It's been plaguing me off and on since 1983, and has mystified doctors until now. It comes and goes, is not contagious, will not kill me. Like canker sores, hives, and other painful ephemera, it's a benign but annoying autoimmune disorder, something you die *with* not *from*. The cause is unknown. It decamps as mysteriously as it arrives, and years can pass between visits. My immune system may be confusing friendly mouth bacteria with foe. Stress can produce an outbreak. Most of the time it's a sleeping beehive. A friend offered: "What happened last week made you lose face, stung you ... and then your face began to hurt." Too metaphorical? Not in the wonderland of brain logic. The mind can be devilishly symbolic and use the immune system as pain-artist, the body as canvas. Whatever the cause, my immune system is shadowboxing. I'm lucky nothing systemic is raging, for the immune system can also sponsor a host of severe diseases, including lupus, rheumatoid arthritis, and diabetes. Is the brain part of the immune system, or is the immune system part of the brain? They're in cahoots, merciless allies, and they rarely forget a face.

Immune systems carry a grudge. When special immune system cells find bacteria, fungi, viruses, or other invaders, they collect them and take them to one of the thousands of lymph nodes scattered around the body. There T-helper cells receive the cargo and order B cells to manufacture antibodies, proteins that stick to the invaders and kill them. Other immune cells save pieces of the invaders as memory aids. They keep mumbling about the invader, and the next time it appears, the mumbling surges to an all-out war cry. Only this time the immune system fights harder. Our lymph nodes keep a roll call of each enemy—every flu and cold that clobbered us, last summer's pneumonia, the jellyfish stings in the Bahamas, the chicken pox and measles, the protozoan imbibed while snorkeling in the Amazon, and all the inoculations, too. An internal police state, the immune system monitors known troublemakers, and to be safe, all strangers in general. That effort may be imperceptible, but the conscious mind follows suit, while basing decisions (it thinks) on logic. Hence the bottle of soap (made from tea-tree oil, peppermint, and lemon) at my bathroom sink.

Yet as immunologist Gerald N. Callahan observes in *Faith, Madness, and Spontaneous Human Combustion,* we safely trade bits of self with loved ones all the time. Couples pick up some of each other's mannerisms, accents, habits, ideas. But we also absorb people in more visceral ways. When we pass along a flu or cold sore, for instance, viruses pack some of our proteins and lipids in the viral envelope and release them inside another person, who will store some in his or her lymph nodes. Retroviruses—such as AIDS, for instance—can install pieces of someone else's DNA in one's chromosomes. But we're probably swapping gene fragments with people all the time, imperceptibly, through infection and lovemaking because "over the course of an intimate relationship, we collect a lot of pieces of someone else. . . . Until one day what remains is truly and thoroughly a mosaic, a chimera—part man, part woman, part someone, part someone else." Little by little, as bits of DNA make it to our chromosomes, intimate relationships help shape the immune system's cameo of us, and modify the brain, altering the self whose continuity we cherish. We don't just get under each other's skin, we absorb people. Everyone we've ever loved remains with us, and we're invisibly changed for having known them. That will make some people feel queasy, I suppose, but it warms me.

A self can be permanently lost through brain damage. Kill the neurons and you kill the self. But when you try to *lose* yourself on purpose, try to sample another's awareness, you can only go so far. The porch light stays on, the mental stew keeps simmering, even if the home owner steps out briefly or falls asleep. It's nearly impossible to tally a thousand unique selves, let alone the billions of robust personalities with whom we share the planet, sometimes facing off, sometimes oozing with love.

Personality

The meeting of two personalities is like the contact of
two chemical substances. If there is any reaction, both are
transformed.

—Carl Jung

At summer camp in the Poconos, I shared a bunkhouse with ten
thirteen-year-old girls. We had a biological universe in common. On the ledge of life, none of us knew much about sex,
though we talked a lot about it, as we did about crushes and true
love. Would-be sirens, we believed in surefire rules of seduction—
we just didn't know exactly what they were. Algebra and tennis had
rules, and if we'd learned anything from childhood, it was the
inevitable link between cause and effect. Surely snaring a boy's love
was the same? Most of that summer we practiced subtle flirting and
experimented with strategies (talk to him, ignore him, flatter him,
tease him) known only to us. If the boys knew, they didn't say, and
for some reason it never occurred to us that they might be going
through similar maneuvers.

Meanwhile, we learned practical tips about inserting tampons,
building fires, applying makeup, and canoeing. We often gathered
by the lake for hootenannies, and every summer the senior campers
put on a Broadway show. One memorable season it was *Bye Bye
Birdie*.

Some of us took Red Cross Lifesaving, Water Safety Aid, and
Water Survival classes. The instructor, Lou, was a gorgeous college

student from Philadelphia, and we eagerly volunteered to play the role of drowners. The murky lake water made lifesaving challenging, but also tensely erotic. In the underwater approach, the saver dives below the surface well beyond the drowner's reach. When would the touch come? Suddenly I'd feel his hands on my knees, spinning me around, then climbing up my back, then wrapping an arm across my chest as I flailed in feigned protest and he shackled both arms around my chest, scissor-kicking in hard, squidlike bursts as he towed me back to the dock. There he put one of my hands on top of the other on the dock and, holding them in place, hoisted himself out of the water, then reached over to my leg and lifted me out as deadweight. Next came the pretend mouth-to-mouth, though of course our lips never touched. We girls learned quickly and had the muscle power to do underwater approaches, rescues of campers putting up a nasty struggle just for fun, and, hardest of all, pull-leaping from the water and up onto the dock by kicking like a porpoise, while still holding on to the drowner. We didn't know that, in terms of evolution, we were the perfect age to learn such skills *and* obsess about boys. If a biopsychologist had told us that there's a physical basis for teens seeming irrational, we wouldn't have believed her. Teens are tempestuous and impulsive, in part, because the connections in their frontal lobes are still thatching. As brain imaging shows, teens use the amygdala much more to process information than adults, who use the frontal cortex more. So while teens mature, parents often end up serving as a commonsense part of their brain. Of course, we felt reasonable, fully formed, smarter than our parents, bursting with life.

Brains are so plastic during our formative years that if one part is damaged, another can rewire itself and take over some of the work. When we're very young, the brain is even flexible enough that if one whole hemisphere dies, the other can sometimes teach itself to sing both songs.

Though, heaven knows, we threw ourselves at Lou, using all the secret tactics we devised, he didn't succumb. Instead, he seemed smitten with a blond instructor his own age. By fourteen, we were counselors in training, CITs. At home we listened to Connie

Francis singing "Where the Boys Are." At camp, we did exercises
we thought would increase our bust size, and sang this song:

> *Born too late*
> *for you to care for me.*
> *I'm just a CIT*
> *that you can't date.*
> *Why was I born too late?*

We didn't compose it; many CITs sang it before us. Camp tradition
included the reality of teen yearning, which was laughed about but
acknowledged as a normal part of a girl's personality and growth.

Each girl had her own temperament. One might be easily bored
and starved for new experiences, or sex obsessed, or a loudmouth,
or a prankster and clown, or clingy and ingratiating, or a ringleader,
or broody and easily wounded. One girl hated sitting still, bit her
nails, and busied herself to the point of constant self-interruption;
another was pure slugabed and found a sensuousness in sleep that
only two alarm clocks and the horror of missing breakfast could disturb. Though amusing, none of us was shocking, since we were all
familiar types we'd encountered in our families and at school.

Nonetheless, I'm sure we hid a lot from each other. For one
thing, I hid the way my senses mingled, producing a synesthesia I
found luscious, but which seemed to spook people. Some feelings
had tastes; I often thought in sensory images I felt; the world's
details popped out at me. Things reminded me of each other in a
slippery way. One example: a cranky hot water heater sounded like
a giraffe choking on an abacus. On an overcast day, the lake surface
looked like pounded gunmetal. Although I found sensory reci-
procity exciting, I knew the girls would tease me about it, or
worse, ostracize me. I didn't mind their knowing I wrote poetry, but
I hid my regular trysts with the tennis instructor. I didn't meet him
for romance, the basics of which I barely understood, but for the
dangerous, faintly erotic exchange of ideas. He was a philosophy
major at Temple. I hid the borrowed copy of Hermann Hesse's
Siddhartha, which I read by flashlight under the covers at night. I

sensed my intensity would frighten them. That's not to say I wasn't also teenage shallow and often blind to the obvious. It never occurred to me that they might have secret mental worlds, too. Yet despite our different temperaments and idiosyncrasies, we shared most of our mental mannerisms, because humans are distinctive animals.

What characterizes our human personality? How would an alien describe the ways of humans? In 1989, Donald E. Brown created a List of Human Universals, behaviors and traits ethnographers observed in cultures they studied. I find his list an eye-opening glimpse of the quirky humanity we all share. Each of us brings our own exotic traits to these basic themes. The full list runs to many pages. Here are some of the features he noted:

abstraction in speech and thought, actions under self-control distinguished from those not under control, aesthetics, affection expressed and felt, ambivalence, anthropomorphization, art, baby talk, belief in supernatural/religion, beliefs about death, binary cognitive distinctions, body adornment, childbirth customs, childhood fear of strangers, choice making (choosing alternatives), classification (of behaviors, inner states, weather conditions, tools, etc.), collective identities, conflict, conjectural reasoning, containers, cooking, cooperation, copulation normally conducted in privacy, coyness display, crying, culture, customary greetings, dance, daily routines, death rituals, decision making, distinguishing right and wrong, division of labor by sex, dreams, dream interpretation, emotions, empathy, envy, etiquette, explanation, facial expressions, family, fears, figurative speech, fire, folklore, food sharing, attempts to predict the future, generosity admired, gift giving, distinguishing between good and bad, gossip, government, grammar, hairstyles, healing, hospitality, hygiene, in-group and out-group, insulting, interest in living things that resemble us, jokes, kinship statuses, language, law, leaders, logic, lying, magic, marriage, materialism, meal times, meaning, measuring, medicine, memory, mood- or consciousness-altering techniques, mourning, murder proscribed, music, myths, narrative, overestimating objectivity of thought, pain, past/present/future, person (concept of), personal names, planning, play, poetry, possessiveness, practice to improve skills, private inner life, psychological defense mechanisms, rape, rape proscribed, reciprocal exchanges (of labor, goods, or services), rhythm, right-handedness (mainly), rites of passage, rituals, sanctions for crimes, sense of self (distinguished from others, and responsible), sex (attraction, jealousy, modesty, regulation), shelter, social structure, status and roles, sweets preferred, symbolism, taboos, time, tools, triangular awareness, true and false distinguished, turn-taking, tying material (string, etc.), violence, visiting, weapons, world view.

In later years, he added a few more, including:

anticipation, attachment, critical learning periods, fairness, fear of death, habituation, hope, husband older than wife on average, imagery, institutions (organized co-activities), intention, judging others, likes and dislikes, making comparisons, males engage in more coalitional violence, mental maps, moral sentiments, pride, proverbs, risk taking, self-control, self-image (concern for what others think), sex differences in spatial cognition and behavior, shame, thumb sucking, tickling, toys.

What an evocative portrait of a unique animal, the strangest animal on Earth. And it only scratches the surface, ignoring many nuances. Sometimes our science fiction (and our science ignorance) scares us with the possibility of lifelike computers interchangeable with humans. We might be able to program a computer with our combinative skill, train it to factor in the success or failure of past experiences, install a statistical prediction machine, teach it grammar and vocabulary and the likely responses to common utterances. But it would still lack a mind that's complexly human and uniquely personal, one that's brimming with inherited family traits, emotions, the effects of interacting with others, and the random flux of a life. Our human personality is one of a kind, and unlikely ever to exist again in all of creation, on Earth or elsewhere in the universe. Despite our habit of depicting aliens with bilateral symmetry and our jumpy motives, there's been far too much genetic water under the bridge for us to look or act like life-forms on any other planet. That should both humble and thrill us. An important aspect of that human personality is our longing to find the world meaningful, and to note all the people and things emotionally important to us.

Okay, so we share an imposing human personality. And, beyond that, we also have one-of-a-kind personalities. Among many species, individuals vary in temperament, likes, and habits. It's the huge range of variation among humans that's dazzling. How is it possible for vast numbers of humans to have elaborate, novel personalities?

Other primates can walk on two feet for short spells, though not comfortably for long. Lots of animals are bipedal. Birds, kangaroos, and dinosaurs, to name only three. When humans began walking upright as a way of life, hips became thicker to support the weighty head and torso, and that forced the birth canal to shrink. Like other primates, we gave birth to well-developed newborns, but our brain was growing too large to bear, and we ran the risk of most babies dying, our species going out like a flame (the etymology of *extinct*). The solution we found was to give birth earlier to a baby with an unfinished brain only about 25 percent of its adult weight. By comparison, a newborn baboon's brain is about 70 percent of its adult weight, and within a day or so it can cling to mother. Most of what a baboon baby knows it inherits, arriving rich with instincts and a largely formed brain, ready to learn the rest. We're born completely helpless, more at risk. And what a feast of surprises awaits us. As the playwright Christopher Fry puts it in *A Yard of Sun*, we're "Dropped into life, after so long in a place / Where life was like a belief in the supernatural."

Human babies fall into the world headfirst, and can't survive on their own for years, while the handyman brain finishes wiring itself and a quirky self ripens. Although family genes play a big role, humans evolve wildly different personalities because the brain does much of its growing outside the womb. A newborn's brain contains billions of neurons, but many of those aren't complete yet, and the cerebral cortex isn't ready to do its dance. At birth, the brain's connections are few, but it builds them at a furious pace until the age of six, when they're at their densest. The brain sensibly bushes out before the violent topiary work begins. Then the connections are severely pruned, some are strengthened and some discarded, to sculpt a fitting brain that also fits the skull. What gifts may be tossed on the trash heap during that savage cut? Infants experience synesthesia, which most people lose as they grow up, but not all. Babies find sensations gushing in higgledy-piggledy. As the cortex matures, it learns to filter and sort, directing information to different locales. Because a flood of sensations could jam up consciousness when we need to act, the developing brain becomes more discriminating, more exclusive.

What ills might the brain's pruning errors unleash? Do they play a role in some forms of retardation and help explain idiots savants? Essential brain parts operate at birth. The brain stem regulates the body; the cerebellum moves it; the thalamus feeds it sensations. The rest takes years to finish, producing the longest infancy of all mammals.

During that spell, the brain is crafted by genes, family, and experience, a mix that's different for every child. As the NYU neuroscientist Joseph LeDoux puts it: "Whether your paycheck is deposited to your bank automatically or you hand it over to the teller in person, it goes to the same place." Nature and nurture "are simply two different ways of making deposits in the brain's synaptic ledgers." Our brain's genetic plan ordains that it reprogram itself in response to its surroundings—including the womb environment, where the tidal landscape is mother, the nourishment her food, the weather her moods.

Many adult dispositions and diseases may well begin in womb time, that lost world of tropic heat and seething storms, where we age before we even begin. In that nursery bay, a mother's poor diet or roiled hormones (caused by fear, depression, or stress) can alter the developing kidney, liver, or brain. Especially, grossly and subtly, the brain, which inherits its basic plan but sculpts important landmarks and pathways while in the womb. According to the epidemiologist David Barker, of the University of Southampton, poor nutrition or infection in the womb and just after birth can program the developing heart, liver, pancreas, brain, and other organs for a host of diseases in adult life. For example, he finds that babies born to undernourished mothers are more likely to become obese adults, perhaps because they're prepared in the womb for a meager life with scant nutrients, and face instead a diet high in fat, sugar, and calories. Not everyone agrees with Barker's theory, but it makes sense to many. What's scary, of course, is research suggesting the colossal damage that even small shortages of protein or certain vitamins can exact. This sort of thinking can produce a fatalism that's unnerving. There's no doubt that heavy smoking or drinking during pregnancy can impair a developing fetus, but how about the mother's mood? She may not have a good diet, and may be angry and depressed every day. In theory, that could spike the womb enough to influence the development of organs, blood vessels, and neurotransmitters. So where does personality reside? Is it all genetic, blended in the womb, or acquired as one lives? Surely the correct answer is all three.

Influenced by genes and the sway of womb time, human newborns arrive with budding personalities. As twin studies have shown, much of one's personality, including penchants you'd think were arbitrary (how one parts one's hair, color and type of car one drives, political leanings, and so forth), seems to be inherited. I mean a cast of mind is inherited, not a specific yen for a Toyota. One's nervous system could easily predispose him to like Jackson Pollock's paintings, Thai cuisine, and riding motorcycles. Another's to recoil from orange juice, find Heitor Villa-Lobos's *Bachianas Brazileiras* thrilling, and flee winter in Minnesota. What's surprising is how much twins can differ. Despite their identical genes, if one twin has a mental illness, the odds of the other developing it are only about 50 percent. The brain refines an elaborate self in response to its genes, early caregivers, friends, birth order, experiences, and culture—all of which provide lessons, trials, and expectations. We usually think of choice as freeing, as widening one's scope. But every choice also narrows, because it rejects a world of competing options, however rich. The brain wiring we inherit can partly rewire itself at the drop of a hat or a disappointment. All learning leaves traces. The brain has a vested interest in pretending that it doesn't, so that the world will feel solid, safe, and predictable. Not a careening carousel one clings to for dear life.

In the loam of childhood, sparked by genes, a detailed self begins to rise. Through "complex and subtle negotiations," the psychoanalyst Stephen Mitchell writes, "the infant and the mother mutually shape each other to create a world into which the growing child will fit." The mother provides her own special emotional climate and sensory universe, where a child learns what to expect and begins to interpret self, others, and world. "My own personal research," says the psychiatrist Allan N. Shore, of UCLA School of Medicine, "indicates that the orbital prefrontal areas undergo a critical period of growth at the end of the first and into the second year of infancy, and that extensive experience with an affectively misattuned primary caregiver creates a growth-inhibiting environment for a maturing corticolimbic system . . . a predisposition to later psychiatric and psychosomatic psychopathologies." Because they're a brand of learning, life's traumas and tragedies physically alter the brain, especially during one's first ten years; sometimes we're haunted by them throughout our lives. But at any age, when we experience things, the brain creates feelings, which in turn alter brain chemistry, which in turn produces mental states, which in turn generate feelings, which in turn alter brain chemistry. Social evolution is a powerful process. But so is heredity. Some aspects of personality come with the genetic suit. How we think about ourself in relation to others modifies and redefines that self. Dicey as that strategy may sound, it gave our ancestors a priceless tool for survival: variety. If humans could just survive infancy, they'd encounter a landslide of experiences, insights, and know-how. Variety is a triple-edged sword. It's evolution's treasure, but also the source of much loneliness, alienation, and misery. Glorious art couldn't exist without that variety, but artists tend, rightly, to feel marginal.

With helpless newborns to feed and protect, we became deeply social and even more dependent on each other. Because our histories differed, we acquired specialities. Both skilled and flexible, we swapped information and expertise. Soon we learned that by pooling knowledge we could enlarge our scope and enter new habitats. Unlike other animals, we adapted to a startling range of landscapes, hardships and climates. Being born with a large, immature brain was advantageous, but it also left us with an urgent need for others, a social bond so compelling that we suffer tragically when it's broken and panic if it's threatened. Humans cooperate far more than other animals, something our brain rewards us for by activating the same pathways stimulated by chocolate and cocaine. We tend to cooperate because we think we'll be rewarded for it, externally or internally, and we're right.

Unlike cats and dogs, we don't give birth to a litter but create very few young, rarely more than one a year. If that one dies, there may be few backups. Thus monarchs are traditionally encouraged to have two sons, "an heir and a spare," as the saying goes. To ensure their offspring survived to build families of their own, our ancestors had to make many decisions based on the obstacles and threats they encountered, some predictable, traditional even, others unforeseen. That required a subtle and flexible brain, a brain steered by instincts and reflexes but also good at improvisation, full of swerve and cunning, a brain that welcomed novelty.

Since both individuals and their adventures varied, they evolved personal strategies, emotions, beliefs, habits, preferences. This combination of rigid behavior on the one hand, and adaptability on the other, is why all people are alike but everyone is different. We call this *personality,* and we say it is something one develops, as if it were a photographic image emerging from the darkroom of one's past. How natural, and in many ways how animal of us. In rapidly changing and unforgiving landscapes, the animal with the best chance of survival can detect new experiences, quickly review its options, decide what to do, and learn from its choices. Flexibility was and still is our genius. We are gifted generalists. We sample. We analyze. We learn. We form opinions. We change our minds. We avoid danger. We bend to pressure. We persuade others. We are persuaded. We take risks. However, Sophocles was right: "Nothing vast enters the life of mortals without a curse." The more we've out-smarted the environment by changing our way of doing things—designing antibiotics, for instance, or building houses with chimneys—the more we've created new problems (resistant bugs, pollutants) that must be solved.

As with most things, genes influence but do not determine personality. A bad gene may be offset by good ones, a bad gene may be mitigated (though not erased) by love and care. For example, the psychologist Avshalom Caspi and his colleagues at King's College in London conducted a long-term study of maltreated adolescent boys, following them up to the age of twenty-six. They found that boys who inherited an active version of the gene MAOA were less likely to engage in violent crimes and antisocial behavior than boys born with a mellower version of the gene. Most of the mellow-gene boys became bullies, thieves, violent offenders, and felt little remorse. Of the 442 abused boys, only 12 percent had the dangerous version of the gene, but they later committed nearly half of all the violence. Abused young men with highly active MAOA genes tended not to become criminal, despite their abuse. That may seem to argue for genes fixing one's personality. What's fascinating about this study, in which childhood abuse *in combination* with a gene led to trouble, is that it shows how upbringing and genes interact to produce behavior. Two thirds of human beings don't have the high-risk gene, which may help explain why some people endure accidents and traumas better than others.

Another study at King's College showed how long and short variants of a gene regulating serotonin levels seem to provide risk factors for depression. Terrie Moffitt and her team studied the nor-mal meteor shower of stresses experienced over five years by 847 white New Zealanders, whether they became depressed, and which version of the serotonin gene they happened to carry. People with two long copies of the gene were able to bounce back, even after traumatic events. Those with one long and one short copy had an increased risk of depression, and people with two short versions were prey to severe depression. A lucky 30 percent had two long forms, the lottery double that made turmoil easier to endure; 50

percent balanced one long and one short form; and 20 percent, the group most depressed, had two short forms. How many such vulnerability genes do we host? Swarms, I presume.

We evolved full of fight and courage, in deadly landscapes, where we hunted food and defied predators. Our bodies still quiver at the thought. Danger excites all the senses, a heady feeling people enjoy when it's pretend and in small doses—say, in books and movies. Most people are happy to bridle more dangerous hungers, or redirect them into jobs, hobbies, and sports, where loss of life is less worrisome than loss of self. But for others, a predictable, well-behaved life feels like being suffocated by a slowly squeezing python.

Craving pulse-revving moments, some people pursue an intense life full of conflict and variety—either mental or physical, sometimes both. Mental risk takers may be brilliant, manual laborers, self-taught, artistic, scientific, business savvy, criminal—even a combination of all of the preceding. Leonardo da Vinci comes to mind (it was criminal to be gay in sixteenth-century Italy, as was stealing corpses for anatomy studies). We picture thrill seekers as daredevils, adventurers, and scofflaws, a crazy tribe of misfits that includes spies, round-the-world balloonists, and juvenile delinquents. Those born with what's sometimes called a type T personality tend to have unusually flexible minds that crave stimulation. They tackle problems from many angles, relish offbeat solutions, skip easily between the known and unknown, enjoy juggling the particular with the abstract, and welcome ambiguity and paradox. They love to break rules, overturn tradition, buck authority, explore intense sensations and emotions. Thrill seekers tend to be astonishingly creative—or destructive—and more men than women fall into the category, especially young men between the ages of sixteen and twenty-four. I know some such people who began by taking physical risks and switched to less lethal intellectual ones as they aged.

What creates a risk-taking personality? It mainly comes with the genetic suit. People with nervous systems that respond drably to stimuli can feel too calm for comfort, and crave a boost of adrenaline and dramatic events to rouse themselves. Two independent studies reported in *Nature Genetics* found that people given to "novelty seeking" had an unusual version of D4DR, a gene involved with the brain's response to dopamine. This echoes other research linking dopamine and novelty seeking. However, personality depends on a blend of genetic factors, and in any case, genetics alone can't completely forecast someone's personality, which upbringing, culture, and life experience uniquely season.

So are we stepchildren of nature or nurture? Even to ask that question implies a dichotomy nature doesn't pose. Only we pose it. It's easier for our brain to handle alternatives, to divide every issue into extremes, which requires less brainwork to fathom and less time to evaluate. In the ancestral world of predator and prey, time was risk. So we impose artificial extremes on the world, but the world isn't that unsubtle. As the social psychologist Deborah Tannen argues so well in *The Argument Culture*, life rarely offers clear alternatives. Most of life sprawls on a continuum of possibilities, compromises, extenuating circumstances. An example of our bias is the benighted notion that opposite views must be offered in the media, for the sake of fairness. Authors find this especially annoying because if there are a hundred positive reviews of a book and one negative review, reference books, online bookstores, and the media offer one positive review and one negative, which suggests that half the critics hated the book. In the media, balanced means positive plus negative, regardless of the ratio. A more disturbing example is television news shows, which also feel that balanced reporting means providing two extreme views. Yet most of anything falls between opposite poles, every idea or feeling includes gradations. Although it is the way of our kind to seek pattern and simplicity whenever possible, that's not something we should honor when complex issues are at stake. Maybe we don't need political candidates to hold opposing views on everything. Maybe they're appealing to our brain's tendency to coast, laze when not threatened, save our detailed scrutiny for more passionate concerns, such as the revised offside rule in women's soccer.

Nature and nurture aren't rivals, nor are they conjoined twins separated during a marathon operation. Both contain innumerable fates and processes, jostling membranes too dense, iffy, and contingent to pin down from one microsecond to the next. Given that divine flux, forget about truthful predictions. *Useful* predictions is all the brain requires. Its reality isn't a monk seal's. It doesn't know real or even truly real or even absolute from otherly real. Aiming for objectivity is its way of feeling a little less subjective. But true objectivity would mean standing outside the human body, off the Earth even, observing both without bias and without a human brain. Of course, that viewer would entertain its own subjectivity, if it thought in such terms, which I doubt. I think, therefore I attribute thought to others.

Empathy, rightly our pride and joy, hinges on an ability to analyze ourselves and assume others must feel the same. Then we can predict how our actions will influence theirs, and vice versa. Of course, they're being affected by others more influential than we, and an inner ecosystem denser than the thickest jungle, and a salvo of self-stimulating thoughts, and gene variants they wouldn't understand anyway, not to mention a lifetime of emotionally charged memories. Still, we have the chutzpah to try. But if we had to think all that through every time we asked someone for a sheet of stamps or said "I love you," we'd grind to a halt in an entanglement of thoughts I personally would love to overhear or read between the covers of a book, but some might find impractical for everyday concerns. The same thing would happen if I tried to plan the grammar of a sentence like that last one. Little of our mindplay hits conscious awareness. And little of our body's doings. Rarely do we note the cat's-whisker delicacy we rely on to sense a snowflake landing on our cheek, or know if we're dressed. For the most part we ignore the body's toil, unless something suffers, breaks, or goes wrong. If we were more aware of our physical selves, we'd tire out fast in an avalanche of sensation.

Ultimately, personality springs from both nature and nurture: experiences of all sorts and our genetic heritage. The pair influence everything, including deeply personal, idiosyncratic traits (some inherited, some shaped by family life, some thanks to the vulnerabilities all humans share) with more or less severity, along a vast continuum. Shake that bag of tricks and today's personality emerges. Tomorrow it will differ a little, depending on sleep, work, food, cranial-sacral therapy, sex, fresh air, a virus's onslaught, prolonged periods of cuddling, a surprise in the mail, or just an innate streak of perversity. One needs both the genes and the gee-whiz to feel curious. An emotional traffic accident would change that fast. If you have the gene that protects you from a spree of stress hormones, the damage might be minimal. That our genes determine talents, temperaments, general intelligence, and susceptibility to alcoholism, depression, and other diseases, we've known for some while. But how traits become expressed, or ameliorated, or stressed, or inhibited, or promoted, or sublimated, or warped by life could fill libraries—and does.

We vary in many predictable ways, especially these five: extroversion or introversion, antagonism or agreeableness, conscientiousness, neuroticism, and openness to experience. Those well-studied personality traits help explain styles of coping and behavior. But they're only the scaffolding of personality. For example, being antagonistic and burly doesn't mean you're going to be a great football player. I know some antagonistic burly writers (like the man who paid his library fine by pouring a thousand pennies onto the librarian's desk). Genes can't fully explain why a businessman with a high IQ and a flair for math might fail in one business after another, whereas his failing on purpose, thereby forcing his parents to nurture him, because he experienced chronic neglect as a child, might. A desire to make one's parents proud, in terms they can understand, has fueled as many careers as inherited traits. Two people may drive the same car in different ways, depending on training, need, weather, or a passenger's demands. And then there's the question of trauma, which we tend to think of as big bad life events. Trauma can also be subtle, chronic, and perforate or deform naturally inherited traits. As a species, our genetic makeup may indeed be limited and patterns obvious. But because our individual experiences vary, and we have agile brains eager to learn, we become complexly different people.

Still, some people rabidly insist that we're born with our lifetime's personality, that people can't change their hereditary spots. That would only be true if the brain were complete at birth. Children's brains stay plastic for years; otherwise they wouldn't be able to learn a lifetime's language so well. Fortunately, or unfortunately, they're busy learning other lessons too, ones that will also be wired into the brain and may accentuate, modify, or in other ways influence the birth personality. Some well-known theorists have argued that parenting practices don't shape children's personalities much, and if a parent ridicules, undermines, or gives meager affection to a child, little harm is done, since the parents' job is mainly to feed and protect. I guess that's a comforting thought to some parents. However, it's not an attitude shared by most scientists or psychologists, especially child psychologists. Words can be used as blunt instruments. As the King's College study of boys with a genetic predisposition to violence showed, genes *and* upbringing craft personality. Failure to thrive is a well-known syndrome, famously studied among World War II orphans, in which children deprived of touch don't develop normally. Also true for rat pups, even if the stroking is provided by humans using Q-tips. Cuddling children matters.

Why is the skin such an emotional fabric? A light caress, which can feel both ghostly and erotic, ignites its own special nerves. A touch may feel as delicate as the legs of a strolling ladybug, ambiguous, slow, languid, half-arousing. All the while, it's zinging messages to the brain. The neuroscientist Håkan Olausson, of Sahlgrenska University Hospital in Göteborg, was treating an autoimmune patient who'd lost sensation everywhere below her nose, when he found that, surprisingly, she was perfectly able to feel pain and temperature. Only her fast-conducting neurons had been damaged, and that left her without a complete sense of touch, and unable to do the plain geometry of how an elbow or knee is angled in space. Yet when Olausson stroked the back of her hand with a paintbrush, she felt the quintessentially soft pressure. As an MRI revealed, that caress stimulated the insula, a deep brain region that helps process emotions, such as those lovers feel when they gaze into each other's eyes, or a baby feels being cuddled. Caresses subtle as a cat's whisker can incite sexual arousal, triggering the release of vital hormones such as the cuddle chemical, oxytocin. Cuddled infants produce more of it, and become calmer, feed better, grow. When touch-deprived orphans were regularly cuddled by nurses and doctors, they returned to growing normally. Elderly people who have pets live longer. Oxytocin levels soar just before a woman gives birth, and rise significantly after lovemaking (probably why more women than men want to cuddle afterward). Lack of touch may warn the body not to waste much energy growing, since there's no protective parent around.

All learning affects the brain. Mothering affects the brain. So do talk therapy, religion, cognitive behavioral therapy, and teachers. Or, as the neurologist Martha Denckla wryly observes: "Every teacher is a brain surgeon, a scary thought . . . every teacher is making little dendrites sprout and connect up neurons. So we are always training the brain." The more emotional, momentous, or traumatic the learning is, the more ingrained it becomes. Our brains are programmed to learn, to adapt, to be flexible. Neural networks flourish during the brain's first three years, a time of exuberant growth, as the world gushes in and the brain tries to adapt fast and heartily to what it finds. No wonder childhood relationships are powerfully important. They spur growth during this fertile period, shaping the way the brain organizes itself. Believing that we're born with fixed personalities is just as silly as believing we don't inherit much of our essential nature and temperament.

Anyway, which personality do we mean? We're not the same personality with everyone. We adjust our self to each person we meet, each situation we're in. We have a flexible self. In fact, inflexibility of self—fixations, compulsions—we regard as unhealthy. Just as being able to focus hard, but also switch attention, aided our chances of survival, not having to be exactly the same self with everyone makes us more successful socially. Does that feel false? Not true to yourself? Only if you believe in a rigid self that's uniformly on view. If you accept that *self* is a plural noun, more like a repertoire than a statue, then featuring one side more with one friend or associate than another won't seem dishonest. Anyway, we do it automatically, sometimes even mirroring the body language of someone we like in a subtle mambo of gestures.

When people move from one culture to another, they tend to adopt the mental style of the new culture they're immersed in. The psychologist Richard E. Nisbett, of the University of Michigan, has been measuring perceptual changes in people who switch between Western and Eastern cultures. They start sensing objects, timing, and relations differently, as they absorb the fears and goals of their new tribe. Humans are crowd pleasers. Our ancestors learned to behave like the people around them in order to survive, and were agile enough to change often, if need be. In *The Geography of Thought*, Nisbett makes a powerful case for how, over millennia, Asians and Westerners adopted different styles of thinking (Aristotelian vs. Confucian) about society, nature, and self. As a Westerner, for example, I used the idea of *versus* because I've inherited a mode of analysis from ancient Greece, a sea-trade society that favored debate, logic, and categories. The Greeks believed in personal identity, liberty, and goals. Objects (including individuals) were separate from one another, nature was separate from humans, and the best way to understand something was to remove it from its context, then break it into as many separate parts as possible, until you unearthed an underlying principle. Society was atomized, too. A successful person stood out from the crowd and had physical or intellectual power over others.

"For the early Confucians," Nisbett observes, "there can be no me in isolation. . . . I am the totality of roles I live in relation to specific others." Whenever one role changes, it rearranges the others, and so the *me* of a moment ago becomes a subtly different me. In the Confucian universe, the slightest quiver or desire influences everything else in a complex equilibrium where even opposite forces coexist. Because social harmony was prized, self-control became paramount. In place of Aristotle's athletics of abstract thought and knowledge for its own sake, Confucians preferred thought with practical applications. Originality posed problems; a solo self was selfish. Each person's unique temperament and talents added to the flowing mosaic of the group. That compound sense of self was reflected in the vocabulary, which offered many different words for "I," depending on the relationship. How you related to someone determined who you were at that moment.

So, yes, peers and culture are deeply influential. Nature—by which I mean both human nature in general and an individual's unique hereditary bouquet (including gender, life in the womb, accidents, ethnic and family traits)—does contribute hugely to our lives. But nurture also matters. In *The Dependent Gene*, the psychologist David S. Moore, of Pitzer College and Claremont Graduate University in California, offers many examples of traits that *appear* to be hardwired, a gift or curse of one's genes, but really aren't fixed. Instead they arise from the interplay of self and other, when DNA meets world. Children with phenylketonuria, or PKU, for instance, don't make the protein that metabolizes phenylalanine, an amino acid found in meat, bread, milk, eggs, and other foods. Unprocessed phenylalanine jams up the brain, so PKU can produce terrible retardation. Ultimately, doctors realized that children with PKU weren't doomed by an indelible scribble of DNA. Parents and diet can change everything. If the children aren't allowed to eat foods with phenylalanine, they don't suffer from PKU, a disease that's both genetic and environmental.

The world ripples with positive and negative events, which we embrace or flee from, in the process learning, adapting, growing. However visceral those dramas, however tender those relationships, their qualities become codes in the neural networks that sculpt the brain and mind. Experience is translated into neurobiology. The psychotherapist Louis Cozolino sees his calling as "a specific kind of enriched environment designed to enhance the growth of neurons and the integration of neural networks." Extending that line of thought, he believes "what Freud called defenses are ways in which the neural networks have organized in the face of difficulties during development. Defenses are ways in which thoughts, feelings, sensations, and behaviors have been thwarted from becoming integrated within conscious awareness."

Because we're addicted to opposites, and have trouble imagining too large a group of anything in detail, we refer to nature and nurture in shorthand with single words, when each contains multitudes and subtleties, and those voices interact, sometimes cooperating, sometimes warring, sometimes reaching a fragile sense of equilibrium and peace.

Winter makes the wolf of the woods go pale, François Villon writes in his poem "Winter." Looking out my study window, at the bare branches of a Japanese maple tree, I'm surprised by the different shapes snow makes atop branches and twigs: pyramids, mounds, jaws, wings, candles, wedges, and animal shapes—sphinx, gecko, lemur with dangling tail. All arise from the same materials, in almost the same space, and yet evolve radically different forms, influenced by the varied microclimates between twigs and branches. When a breeze purrs over a branch, it splits, slows, becomes ruffled or blocked by the time it finds a second branch, where snow responds to other invisible touches. In a similar way, siblings grow up in different families, with a shifting environment not only influenced by parents and brothers and sisters, but by ricochets from their relationship with each other. Imperceptibly influenced by breezes arriving from the round corners of one's life, one meets equally breeze-swayed others, all moving in a dynamic dance of infinite realignments. Even a minutely different environment is enough to sculpt noticeably different personalities and behaviors, just as it's enough to reform the geometry of snow on a tree limb. No image English offers—say, of dominoes, ricochets, echoes, or ripples—can capture how in the brain everything affects everything else all at once, or how our everythingness always lingers in mind as we engage the everythingness of everyone else. Still, because we've got pattern-mad simplifying prediction machines in our noggins, we assign basic personalities to people, including ourselves. It's the least we can do.

CHAPTER 23

"Shall It Be Male or Female?
Say the Cells"

Shall it be male or female? say the cells,
And drop the plum like fire from the flesh.
—Dylan Thomas,
"If I were tickled by the rub of love"

When I was on French Frigate Shoals, home to the world's
remaining monk seals, we doctored a female whose tail had
been cropped by a shark. She was flown to a Hawaiian seaquarium
for care, because with precious few female monk seals left in the
world, losing even one is tragic. We tend to picture humans as
teeming masses, a single organism quickly swallowing the planet.
But once we were as scarce and endangered as monk seals, a hairsbreadth from extinction. We know this because all the people on
Earth can trace their genes back to a handful of common ancestors.
That's why we're much more closely related to each other than, say,
chimpanzees are to one another. At some ominous stage of evolution, we dwindled to perhaps a hundred humans, and among
those, some would have produced only daughters who didn't survive long enough to reproduce, so their genetic line died out.
Those must have been extraordinary beings, hopeful and resilient,
with a furious life force and enough brains to outwit their enemies
and environment, endure brutal trials, endless indignities, and yet
raise strong children. Why does resiliency surprise us? We are
born survivors.

Evolution plays no favorites; men and women both animate a tribe of genes. Inheritors, we all possess similar-looking brains. Simply because women tend to be physically smaller than men, women's brains are 10 to 15 percent lighter, housing fewer neurons, but those seem to have more connections. That rich connectedness may help explain why women are more prey to depression. Studies show that women ruminate more about emotional things. Some say that women aren't more vulnerable to depression, only more comfortable asking for help. In subtle ways, men's and women's brains are wired differently. About five people in a thousand experience synesthesia, for instance, but over 75 percent of those are women. Females tend to be better at what's now called *multitasking,* a word like a breakfast cereal full of fiber. How about spatially? Show me a simple object and I have trouble drawing its unseen side. Although I have a good visual memory, when I try to picture three-dimensional objects in space my mind's eye goes blank. As a result, it took me longer than male student pilots to learn how to land airplanes, an abstract spatial exercise, especially at night. For ages, it seemed, I landed like tripping over toys in the darkness. I finally managed it in a roundabout way by devising my own private geometry: *when those lines are like that, and the runway is there, and trees and buildings look like that, then I'm at this altitude.* Another good trick: *if the runway seems to be sliding up the window, I'm too low; if it's sliding down the window, I'm too high.* Several things may explain my trouble landing: a personal trait like poor depth perception, cultural indoctrination, or a predilection of the female brain. And a fourth: all of them, mixed differently in everyone. A friend of mine, a nuclear physicist, has a keen spatial and mathematical sense, which she no doubt inherited from her physicist father. But she also grew up in a household buzzing with physics and math, in a college community that fostered such thinking and rewarded it.

Are boys more hostile and competitive by nature, or do we program them to be? The psychologist Turhan Canli, of SUNY-Stonybrook, and his colleagues have been showing emotionally charged pictures to men and women. Viewing the subjects' brains with fMRI (functional magnetic resonance imaging), the researchers found that the women responded more intensely to the emotional scenes, which they also remembered more accurately than the men three weeks later. Canli concludes, "The wiring of emotional experience and the coding of that experience into memory is much more tightly integrated in women than in men." This supports earlier studies showing that, for the most part, women have better autobiographical memories. (Women also have a better sense of smell, which may be relevant.) The study suggests that emotional events may have more meaning for women, who devote more time to thinking about them.

The subjects were asked to identify their response to each picture, on a scale from "not emotionally intense" to "extremely emotionally intense." Some pictures weren't arousing (fire hydrants), others disturbing (mutilated bodies). Shown a gun, the men reported feeling "neutral," but the women reported strong negative emotions. Pictures of dead bodies or people crying also fired strong emotions in women. As did a picture of a dirty toilet, which may at first seem funny but makes sense when you consider the history of women and dirt, the symbolism of bodily purity and keeping a clean house. Emotional pictures stimulated the left amygdala and other brain regions in both men and women, but more areas of women's brains became active. Canli concludes that these "encoding processes may be a mechanism by which women may attain better emotional memory than men." Simply put, women's brains tend to remember emotional events better and longer. This isn't news to most couples. We women are notorious for taking the high ground in fights with a cavalcade of remembered slights. Of course, these differences are just statistical; there are female pilots and male poets. It's far too easy to craft stereotypes from such findings.

In a sardonic New York Times column, Maureen Dowd laments that this is "just the sort of data that misogynists could use to support the old argument that women are too high-strung, thin-skinned and brooding to be trusted as commander in chief. . . . But in fact, all you need to do is look at modern American history to realize that it has been shaped and warped by men worrying about what a cool guy thinks of them. . . . So they can do all the studies and TV shows they want on overemotional and man-crazy women. It is the overemotional and man-crazy men who have messed up American history." Well said.

Some important brain wiring takes place in the uterus, while the fetus is bathed in hormones. A flood of testosterone helps wire a male brain, a flood of estrogen a female brain, inhibiting some features and promoting others. Here are a few of the differences revealed by brain imaging and other research, but keep in mind the caveats *on the average* and *tend to be* with each statement, because individuals vary enormously, based on heredity and the mother's circumstances during pregnancy:

A woman's brain has a larger corpus callosum, the sparkling bridge between the hemispheres, and also a larger anterior commissure, which links the unconscious realms of the hemispheres. This may allow the emotional right side to contribute more intensely to the left side's conversation, thought, and other doings. Men more often focus on a problem with the hemisphere that specializes in it, while women tend to recruit both sides of the brain. As they age, men tend to lose more brain cells in the temporal and frontal lobes, affecting feeling and thinking, while women lose more brain cells in the hippocampus, affecting memory.

In Deborah Tannen's films of children at play, time after time, little girls are overheard emphasizing what they have in common, and little boys roughhousing with each other, or sitting together and talking competitively while avoiding eye contact. In scores of experiments conducted over the years, males do better on math reasoning, figure-ground, and spatial tests, and they have better aim. Females excel at language, social and empathy skills, and spotting the similarities between objects and are more sensitive at hearing and smelling. Males can stomach more physical pain. Women who aren't mothers respond more to pictures of babies than men who aren't fathers.

In a rhyming study, girls and boys were equally skillful, but MRIs revealed that the boys were using only one side of their brain, whereas the girls used both. Reading and rhyming are unisex activities; that they activate different parts of girls' and boys' brains suggests the brains may be organized a little differently. "It may be," Martha Denckla offers, "that the left side of the brain is just very bossy in little girls." For some reason, boys are more vulnerable to learning disabilities such as autism, dyslexia, attention deficit disorder, and hyperactivity. As common wisdom always said, girls develop earlier than boys. Girls tend to sit up sooner than boys, but boys tend to crawl around and explore sooner.

For more than twenty years, the psychologist Ruben Gur, of the University of Pennsylvania, has been studying brain differences between the sexes. In a study of how men and women respond to facial expressions, he found that men have more trouble recognizing sadness on a woman's face than on a man's. Women read all faces much better. "You need to be able to pick up on clues, both verbal and nonverbal clues in your surroundings and your family," Gur explains. "And perhaps that is why women would be more in tune with their emotions." When Gur imaged men's and women's brains at rest, he found more activity in some regions of the men's limbic system. He naturally asked himself about the behavior of other animals with that pattern. "What characterizes them," he concludes, "is that they react to emotional situations through actions. If they are angry, they'll attack. If they are fearful, they'll run away." Women's brains showed more activity in the cingulate gyrus, adjacent to the language areas. "It appears in animals who can communicate, who can deal with emotions in a much more symbolic fashion. We don't need to do very complex research to find that men are more likely to express emotions through physical acts, whereas women talk it over."

In my experience, women tend to worry more about losing attachments, men more about losing face. More men become jealous over sexual infidelity. More women become jealous over emotional infidelity. Could parents and society be embedding those behaviors? Partly so, and that would influence a maturing brain. Although not everyone agrees with evolutionary psychology (we may not have spent millions of years on the savanna; ours may be more of an Ice Age cunning), its explanation for these behaviors goes like this: since each man could impregnate an entire fiefdom with the sperm in just one ejaculate, and a woman has few eggs, it's in his best interests to be promiscuous, and in hers to seek a provider and protector for her child. I think our female ancestors had excellent reasons for being unfaithful, too. Having a backup male to help raise the kids might be useful if her mate died. If her mate became too ill to reproduce, she might still pass on strong genes by dallying with another. Genetic variety might have been an important safety net. If a female bore children by different fathers, each child would inherit slightly different genes, giving her better odds that at least one child would survive. A clever female might cozy up to several males so that they wouldn't hurt or kill her offspring. If the males didn't know who fathered her child, they'd all need to protect it.

Whatever the cause, women with a strong sex drive who were unfaithful to their mates produced more children who survived, and thus the genes for that tendency were passed on. Men and women who felt powerfully devoted to one another as mates also produced more children who survived. Men who impregnated as many women as possible also produced more children, just by sheer numbers, even if they didn't stick around to help raise them. In this way, our contrary sexual urges probably evolved, so that now men and women are happily, gratefully monogamous and yet chronically unfaithful. Or you could argue that society spawns those values. But that still doesn't explain away differences in brain structure.

In a provocative theory spawned by split-brain research, the vascular surgeon Leonard Shlain proposes that our invention of writing created left-brained, misogynistic cultures dominated by a male god. He hypothesizes that "when a critical mass of people within a society acquire literacy, especially alphabetic literacy, left hemispheric modes of thought are reinforced at the expense of right hemispheric ones, which manifests as a decline in the status of images, women's rights, and goddess worship." He doesn't claim writing is solely responsible for the shift from female- to male-centered religions, but an important part of the mix.

What elements characterize a left-brained culture? Shlain argues that "a *holistic, simultaneous, synthetic,* and *concrete* view of the world are the essential characteristics of a feminine outlook; *linear, sequential, reductionist*, and *abstract* thinking defines the masculine." Tracing this motif across history's major cultures and eras, he offers examples of how the daring, risk-taking, bloodthirsty hunter-warrior, ruled by the right hemisphere, subdued the more nurturing and sensitive female in religious and political status, in marriage, and throughout society. The direct cause, he argues, was reading and writing, a left-brain activity. I find his dichotomies and many of his examples a little kooky. After all, writing is a relatively recent invention; there exist even today literate matriarchal societies; male dominance rules many (though not all) mammals in which males grow to be larger and physically stronger than females; the reign of goddess cults at the dawn of human history doesn't necessarily mean that women had higher status then; and so on.

However, he sparks some important questions: Can an entire culture become right- or left-brained? If so, how would one become right dominant, another left? What do such cultures look like? What events or inventions can program the brain in ways that have widespread social consequences? Are we able to rewire our brains— and thus direct our evolution—through the objects we invent and rely on? If so, are we unknowingly sacrificing our human nature to marketeers? For example, because we have visual, novelty-loving brains, we're entranced by electronic media. Over time, can that affect the hemispheric balance of our brain in ways that influence us socially, romantically? A related nutritional question: Does the recent availability of fats, salt, and refined sugars—which our ancestors craved because they encountered them so rarely—alter the chemical broth of the brain in ways that may ultimately change our nature?

The brain determines behavior, but the reverse is also true. Trapped in a bad marriage, anyone's mental and physical health suffers. The unhappily married of both sexes suffer a much higher incidence of everything from heart attack to gum disease. In one study, fighting couples were videotaped, and those who had heart disease and argued with their spouses in a consistently negative way were almost twice as likely to die within the next four years. In another study, a spouse's criticism could increase eye blinking and other symptoms in Parkinson's patients. A fifteen-year Oregon study found that women who felt they had less decision-making power in marriage were more likely to die. During marital discord, women's bodies seem to absorb more hostility and negativity than men's. They're more prey to congestive heart failure, joint pain in rheumatoid arthritis, immune and endocrine system problems, high blood pressure. Especially for women, marital stress can be corrosive. No one knows why.

One's luck may be partly genetic. A long-term study conducted with 202 people at the Erasmus Medical Center in Rotterdam, the Netherlands, identified 18 with a gene variant that buffered them from the ravages of the stress hormone cortisol. As a result, they had lower concentrations of insulin, cholesterol, and glucose in their blood, and less atherosclerosis. Men and women over sixty-five were more likely to have the gene variant, which may confer longevity by protecting the body from diabetes and heart disease. In the MAOA study I mentioned in a previous chapter, although abused boys with a mellow MAOA gene tended toward violent acts, fewer abused girls did, because the gene sits on the X chromosome. Since girls have two X chromosomes, they have a better chance of inheriting at least one of the less dangerous versions of MAOA.

All those extra connections between brain cells may truly pre-dispose women to dwell on a lover's tiff, or relive an emotional memory, worrying it, refining it, ruminating to the point of depression. Women may fret more elaborately about relationships. If that's a curse, it comes with a rich interweaving few would want to lose.

CHAPTER 24

Creating Minds

The truly creative mind in any field is no more than this:
A human creature born abnormally, inhumanely sensitive.
To them a touch is a blow, a sound is a noise, a misfortune
is a tragedy, a joy is an ecstasy, a friend is a lover, a lover is
a god, and failure is death. Add to this cruelly delicate
organism the overpowering necessity to create, create,
create—so that without the creating of music or poetry or
books or buildings or something of meaning, their very
breath is cut off. . . . They must create, must pour out creation. By some strange, unknown, inward urgency they are
not really alive unless they are creating.

—Pearl Buck

Are artists born into a different sensory universe? Originality has
been vital to our evolution, though we've paid dearly for it,
since it marginalizes people and creates terrible loneliness. But it
also leads to art's celebration of unusual sensibilities. Art makes
eccentricity safe. Art offers a refuge from the burden of individuality. After all, "Mozart's *Don Giovanni* sets to sublime music the
life of a lecher and serial rapist who would find no respite in the
courts," the neurobiologist Semir Zeki, of University College,
London, is quick to point out. What if neuroscientists could map
those differences? Artists do, Zeki feels. Documenting the rum-
blings of a quirky brain, the arts teach us about how the brain perceives; they're forms of knowing. In a sense, artists are "neurologists
who unknowingly study the brain with techniques unique to

them," and Zeki envisions a field of neuroesthetics, devoted to studying the neural basis of art.

One is not supposed to think of artists as biologically different, nor of imagination as a chemical fizz. If I see a blackbird perched on a telephone wire squinting down at me with one eye like a rebus, my knowing that sight may be prompted by bits of sodium and potassium sashaying through cells, or synaptic junctions brimming with dopamine, in no way bankrupts the miracle for me. I find it just as arresting and mysterious. But it's only arresting and mysterious because I'm as curious about the sodium as I am about the blackbird, and I inherited excitable senses.

Creativity ripples through my family tree. Born in Russia, my grandfather was a spare-time inventor (of the backless vest and other items). One of his daughters, my aunt Frieda, changed her name to Farita and was possibly the oldest performing belly dancer, right up to her death at eighty-six. Over the years, she just kept adding more veils and performing before audiences with poorer eyesight, in the end giving concerts mainly in nursing homes in New Jersey.

My grandfather's two sons, my uncles Lou and Morris, became electronics inventors. One miniaturized the solid-state pacemaker, and the other always carried a harmonica and could be counted on to organize sing-alongs with strangers at bus stops. My mother, who wanted to be an architect, spent her life crafting and designing things. I know little about my great-grandmother except that, in Europe, she made a living by embroidering vests (only the fronts), and sold sugar cubes to men who drank coffee as they played dominoes. The cousins, on my father's side, became accomplished musicians.

I've always trekked through imaginary worlds, lived on my senses, and fiddled with words. Writing is my form of celebration and prayer, but it is also the way in which I organize and inquire about the world. Driven by an intense, nomadic curiosity, I may find myself in a state of rapture about a field and rapidly coming down with a book. For as long as several years, perhaps, I will be obsessed with the senses, the dark night of the soul, or some other facet of human nature. This creative hunger has not always been a boon. For many years, it alienated me from family and playmates, who found my mental fantasia odd. In kindergarten I was reprimanded for using too many colors to draw a tree's thick, chewy-looking bark. As a college freshman, I flunked Logic.

A classic syllogism goes: "Johnny has a bat. All bats are blue. What color is Johnny's bat?" This is the kind of question that would invoke panic in me. I reasoned like this: "Well, if all bats are blue, and Johnny has a shred of individuality, he'd want his bat to look different. Blue is traditionally the color of sadness, the Virgin Mary, the sky—maybe he'd prefer a color that better reflects his mood or goals. I've noticed that shadows really aren't black, they're blue. Would he want a bat the color of shadow? Blue is a color easily affected by changing light. Do blue bats appear lifeless at dawn, but jewel-like at high noon? Are all bats the same size? Are they crafted of different woods, whose grain might absorb the paint more deeply? What sort of blue is it, anyway—pearly, sapphire, luminescent?" I was altogether too strange to pass Logic.

Mathematics was a language I didn't speak. (Later, as a pilot, I could use a circular slide rule to do computations.) Although my passion is for words, I also love playing with ideas, looking at something from as many sides as possible, lifting up an observation and shaking it to see if a revelation might fall out. I don't regard that as work, though it does require ladles of energy, but rather as a fun form of mental mischief. Not ha-ha play, but what I call deep play, a flow state when all the rules of play are present, but raised to transcendent and richly fulfilling heights. Like most artists, I don't want to be creative all the time (otherwise how would I deal with the UPS man or buy a bicycle pump?), but somehow creativity suffuses each day.

Every artist's rich style and eccentricity lives somewhere in the cerebral cortex. Sensory details fascinate me, and always have. I was born that way. I enjoy life's panorama and complexity, a pleasure fed by chemicals and electricity, some of which I inherited, some of which emerged from experience. I was born a sensualist. I bet more people are than stay that way, as the brain tinkers with its

wiring. And it may even have something to do with all the hormones flooding my mother's system when she was pregnant with
me and in daily domestic wars with my dad and his mother, who
lived with them for the first ten years of marriage. It may have
something to do with how my genetically unique brain learned
from and coped with early childhood, and then the spree of life
with its assorted joys, traumas, accidents, and other influences, not
forgetting the DNA I absorbed from intimate others along the way.

I'm a synesthete, but perhaps not as harlequin as the novelist
Vladimir Nabokov, who indulges his "colored hearing" in *Speak,
Memory*:

> The long "aaa" of the English alphabet has for me the tint of weath-
> ered wood, but a French "a" evokes polished ebony. This black group
> [of sound] includes hard "g" (vulcanized rubber); and "r" (a sooty rag
> being ripped). Oatmeal "n," noodle-limp "l," and the ivory-backed
> handmirror of "o" take care of the whites. I am puzzled by my
> French "on" which I see as the brimming surface-tension of alcohol
> in a small glass. Passing on to the blue group there is the steely "x,"
> thundercloud "z," and buckle-berry "k." Since a subtle interaction
> exists between sound and shape, I see "q" as browner than "k," while
> "s" is not the light blue of "c," but a curious mixture of azure and
> mother-of-pearl.

Nabokov loved image-knots, coded word games, pun-slinging,
the voluptuousness of phrasemaking, the way sentences snake
around the mind. In his novel *Ada*, characters pun in five languages,
which he doesn't bother to gloss. Such a literate prankster might be
blessed with a sort of synesthesia on demand, rather than its more
annoying extremes, which afflict some people. "The pen is the
tongue of the mind," the novelist Cervantes writes in *Don Quixote*.
An easily accessible synesthesia, relished by a host of artists, from
the poet Dylan Thomas to the composer Nikolai Rimsky-Kor-
sakov, may be a remnant of infancy left in our psychic wardrobes,
one of the brain's delightful oversights. Thanks to a pruning error
or a genetic mutation, one region of the brain can exchange views

with a neighbor it would normally ignore. That makes it combine unrelated things—the work of metaphor—and indeed synesthesia occurs seven times more often among artists.* To delve into that sensual lushness, one dulls the brain's sorters temporarily. Hallucinogenic drugs work, too, of course. Some psychedelic drugs bind to the serotonin receptors we use while dreaming, allowing one a waking dream state. When I was a college sophomore in the seventies, hallucinogens like LSD, mescaline, and opium were the campus drugs of choice. Out of curiosity I sampled them and found them unexpectedly boring. Yes, a red curtain might become an American eagle flapping across the wall, and the glowing embers in a pipe might squirm like bees. But I could shuffle my senses like that on purpose, especially while writing, minus the side effects, without losing control. Those sorts of drugs limited my sensual scope rather than expanding it, and I bet that's true for many other synesthetes.

Creative ideas are forged in an alchemy of mind, as the brain uses electrochemistry to confect ideas, and then more electrochemistry to think about those ideas, and so on in an endless hall of mirrors. This rarely happens in a tidy sequence. The brain can hold an idea in its stockroom for years, occasionally checking to see if it has changed at all, revising it a little, and then putting it back on the shelf, taking it down again when it seems to have evolved like a lemur from its original form. Then, the first version of an idea might look vanishingly small in thought's mirror, where the most recent reflection looms largest.

Vital to all of our endeavors, memory really stars in creativity. How does memory function in art? Here's my own experience. I'm usually working simultaneously on a few projects, and they may be in different genres. But I think of my mind as a sort of captain's desk with many drawers. When one drawer is open it has my full attention and I'm not aware of the other drawers. After a few minutes or a few hours, I might close that drawer and open a different one,

---

* I discuss synesthesia and some famous synesthetes at length in *A Natural History of the Senses*.

which in turn will have my full attention. What I'm not able to do is close all the drawers, forget what's in them, and walk away in a comfortable amnesia. What I'm working on is always nagging at me, always on my mind, waiting to magnetize daily events to the purpose of the work. It is a kind of remembering. *Keep your mind's eye open for relevant morsels,* it seems to say. *Be vigilant, be ready.* In that way, I tend to see life through the lens of the book I'm writing. The prose books I write are feats of comprehensive remembrance aided by research. While I'm writing them, every observation, every news report, every conversation with a friend seems to illuminate a dark corner of the human condition.

Another kind of remembering is finding my way back to the emotional locale of the book when I want to work on it. Many writers use memory aids to speed this journey—listening to music, taking drugs, reading passages from favorite authors, and so on. I'm not the sort who needs a trail of bread crumbs to find her way back to the campsite, because I don't stray far from the fire. I keep the coals of the book hot for the entire writing and production spells. I may fan those coals to flame on days when I'm writing, but I never allow them to cool until it's safe to, or, to put it more mundanely, I don't allow the intense focus and enthusiasm of the book to lessen until I'm positive it is really finished and I no longer need to be so enraptured. It is a form of extreme attentiveness. Then I sometimes actually hear what sounds like organ stops, or a heavy door slowly closing. In the exhaustion that follows (which for some reason I always construe as laziness), I exhale for what seems ages, and I'm greatly relieved to have my life back. The world ceases to be slanted in one direction. Nothing has to add up. I don't have to concentrate so hard. I don't have to muscle into anything. It all feels like too much bother. The thought of writing anything makes me nauseous. I wallow in fatigue and the delusion of normalcy.

Only after twenty-some books have I come to recognize this state as one of necessary rest and repair. I'm the sort of enthusiast who would lift weights every day of the week to achieve her goal faster; but I realize that you can only gain stronger muscles and bones by including rest in your regimen. I've learned that it is possible for me to sprain my emotions. I have to cut myself a little slack, and allow myself not to start a new book for months on end, if need be. The euphemism I use is "letting the well fill up." Translation: paying less conscious attention, while acquiring new memories and associations.

I'm usually happy writing poems during this period, because I can work on them in fits and starts. Writing poetry requires a different kind of remembering. One makes metaphors and similes by relating something in the present to something elsewhere or elsewhen. Even though I have a poor practical memory—I'm not good at remembering an event years ago or the exact words someone just used in a conversation—I have a generous visual memory. When I look at something, I quickly remember related things. I might see the wind blowing leaves across the lawn and remember a waiter in a posh restaurant whisking the crumbs from the tablecloth after a meal. I might see dry seedpods hanging on distant trees and, although I can't hear them, remember the sound of tiny gourds being shaken. That form of remembering fills me with incontestable joy. A wordless joy. That's why some of my greatest pleasure and satisfaction comes from writing poetry, and why the sections of any book that I'm happiest with are the more poetic ones. However these forms of memory may differ, they rely on the same molecular processes, but they may involve different circuits. For example, they may involve the natural brain process of "spreading activation." When we hear a lot of words, we busily translate them, exciting an area of the cortex where their concepts are stored, which excites a lot of pathways leading to related concepts and words. Some people do this visually as well as verbally, but most do it automatically, unconsciously. Maybe some of us do it semiconsciously.

We may think of artists as supremely self-centered, narcissistic egomaniacs whose subjectivity feeds them—and we'd be right—but they're also inheritors of influential public and historic memories having little to do with them. Every creative person, Jung reminds us, is a bouquet of contradictions. "On the one side he is a human being with a personal life, while on the other side he is an impersonal, creative process. . . . As a human being he may have moods and a will and personal aims, but as an artist he . . . carries and shapes the unconscious, psychic life of mankind."

So are artists born into a different sensory universe? I think so. I don't know why it happens. It may be hereditary, or compensatory, or an incidental by-product of a brain anomaly. Artists and other creators also develop startling expertise, which probably means they're superbly good at paying attention, or, as I prefer to think of it: the useful application of obsession. They also have minds that can obsess uselessly, slum, stall, narrow, issue pigheaded *non serviams*, regress to embarrassing hellholes, and weave taffy-like yards of bland chitchat. Minds do. But I feel that's beside the point. Minds tend to be sloppy and bone-idle, even the best minds. They yearn for entropy, and in time find it, alas. That they can rouse themselves with spanks of imagination and insight is truly dazzling. A key talent creative minds share is being able to concentrate happily, narrowly, for long spells. It's like throwing a lavish party in only one room of a house. Interrupt that intense focus, and the person may seem addled at first. It doesn't matter if she was creating using words; switching to speech and social protocols means abruptly changing mental states, and that can feel confusing. Sometimes the trick is to hold the creative state, with all its pyramidal weight, while simultaneously holding a conflicting state—say, a brief excursion into sociability.

Dispositions can be inherited, just as one inherits handedness or eye color. Inheriting a talent doesn't insure that one will use it, but it does raise the likelihood. Creative gift doesn't start as a memory skill but as a mode of cognitive processing. One person is more visual, another better at abstract thought. The memories they embed reflect those specialties. Monet, with his passion for water, light, and garden, savored subtleties in them most of us miss. A man may be born with a musical ear or a silver spoonerism in his mouth, but expertise in music, literature—or surgery, for that matter—is created through use. Famous experiments with chess players showed they could glance at a chessboard for only a few seconds and later remember where every piece stood. But if they were shown chessboards whose pieces were randomly arranged, memory stumbled. It wasn't a flawless memory that gave them the edge, but familiarity. Because they'd played games and remembered boards thousands of times before, they could "chunk" information into meaningful groups that were easier to recall. As a result, they parsed a board quickly. When I was little, my uncle used to amaze me by rattling off a long list of numbers, which I would dutifully write down. Then he'd recite them either backwards or forwards. In time, he revealed his trick—they were subway stops he'd learned over the years. Chunking the numbers made them easy to remember. What we like to do becomes the thing we do often, and the thing we often do becomes the thing we do best.

However, expertise contains a trap. Familiarity reduces the thrill. *Oh, that again,* the mind says with a yawn, *another wing walker, another balloon crossing.* Mastery may be what we strive for, but when we achieve it we lose the novelty, innocence, tension, striving for accomplishment, and all the other attributes of a satisfying challenge that's only a hairsbreadth more doable than it isn't. We lose enthusiasm, our possession by the gods. The same drug doesn't springboard us to the same heights. For some people, that means increasing the dose, raising the stakes, making the problem thornier, lifting the hurdle a notch higher, doubling the wager, choosing a more dangerous climb, attempting an even more devilish piano sonata. When we become too good at something, we start to take it for granted and then slur over the details. That's why I love learning something new; then every sensation sparkles.

These days, I delight in being able to create my own astonishment, to study the intimate life of a wren family, the shape of rain, or how the brain becomes the mind. For the most part, I can follow my curiosity wherever it leads, and that makes me very lucky indeed. Writing this book is a great adventure and a mystery trip, as is each book. In this one, I combine my favorite fascinations—nature and human nature. On the days I don't work on it, I feel it squirming and tugging and know I'll have to start attending to it again soon. Writing, after all, is an art form that lives along the vertebrae and pierces the heart. A book can clamor for attention.

All humans are creative, some are just more or less so than

others. In fact, I believe creativity is our ecological niche. So let's consider highly creative people in another field. Mathematicians often use the term *beautiful* when they describe a wonderful solution. Ask them what they mean by that and you won't be treated to a beautiful reply. It's hard to define, but easy to recognize. I think those worthies may be responding to the way complexity excites the mind and order rewards it. I think they may mean that when a simple, irrefutable solution can be divined from a thorny problem that has bested brilliant minds for ages, they experience a feeling of blended awe and relief, a state of vigorous calm, a comforting thrill. Their answer fits perfectly into the puzzle box, and when it does, something greater than its pieces emerges.

For the mathematician Henri Poincaré, beauty was classical, not baroque, a simple purity of line and idea, the absence of ambiguity. Nature doesn't always reveal itself that simply, but when it does, it strikes a mental chord. Doing creative mathematics meant discovering the rare "unsuspected kinship between . . . facts, long known, but wrongly believed to be strangers to one another." For Poincaré, to invent was to choose. He remembered vividly when the solution to a particularly stubborn problem came to him:

> Just at this time I left Caen, where I was then living, to go on a geological excursion under the auspices of the school of mines. The changes of travel made me forget my mathematical work. Having reached Coutances, we entered an omnibus to go some place or other. At the moment when I put my foot on the step the idea came to me, without anything in my former thoughts seeming to have paved the way for it, that the transformations I had used to define the Fuchsian functions were identical with those of non-Euclidean geometry.

Where did the sudden illumination come from? The subliminal self, including long spells of unconscious toil beforehand, priming the pump, laying down a bedrock of memories. And afterward, the work of shaping, deducing, verifying. Any great lightning-bolt insight requires a before and after. But significantly, Poincaré observed, "It is not merely a question of applying rules, of making the most combinations possible according to certain fixed laws. . . . The true work of the inventor consists in choosing among these combinations so as to eliminate the useless ones or rather to avoid the trouble of making them, and the rules which must guide this choice are extremely fine and delicate . . . they are felt rather than formulated." The *aha!* moment happens when they're allowed into consciousness.

He imagined a sieve, an esthetic sieve, with which he panned for golden laws. A fetching solution must be both beautiful and useful. Also true for great improvisation in the arts. The trick is not to choose the first thing that pops into your head, but the *best*. That requires years of technical mastery and learning what's possible, as well as freely flowing ideas, and the willingness to take risks, among other qualities. The British landscape architect, Gertrude Jekyll said she used a mental sieve "through which to pass many matters in order to separate the husk from the grain."

I've noticed that the image of a sieve frequently occurs among creative people, who feel their gift doesn't lie in hatching copious solutions to problems, but in being able to apply an extra filter that will let only the best ones pass through. This isn't just a phantom human feeling, but one of nature's basic techniques. When plants photosynthesize, they send nutrients (a flood of sugars) down to the roots, moving them from one cell to the next through microscopic *sieves*. We call them sieves, and even invent household and industrial sieves we can hold, to complement the sievelike processes we can only feel and imagine.

World is all sensation, more than we can register in a lifetime, let alone a moment. Rather than drown in a sea of incoming information, the body uses assorted sieves. The hypothalamus filters sensory news from the body to the brain. The caudate nucleus, lying beneath the cerebral cortex, is thought to filter extraneous impulses and thoughts—some think it may malfunction in people with obsessive-compulsive disorder, allowing in thoughts and impulses usually filtered out. One needs a mental sieve, and fortunately the brain grows deft at filtering and sorting a waterfall of fascinating or delectable details that it judges unnecessary for survival, that is, for life in the present and future. The body encounters a chaotic universe of sensations, and sends only what seems most meaningful to the brain. What will it be today, watching a frog until one notices that it can't swallow with its eyes open? Maybe dowsers slink into altered meditative states in which life's sensory flow recedes and they can read subtly charged electromagnetic fields produced by water moving underground.

When creative minds invent something new, they add thrillingly (sometimes alarmingly) to the tincture of sensory information that passes for everyday life. They force the filter to open wider. What they add becomes as real a fact as the sun. To some extent every thinker or artist is like an idiot savant. Dominating the psyche, a gift may shine like a beacon and the rest of the mind be quite juvenile, squalid, ordinary, a wen of tackiness and self-regard. One may even require a mundane life to accommodate the weight of the gift. Sometimes it's a rude shock, for example, to meet an artist and discover that the public self was the best the person could rise to, in privileged moments, whereas the off-duty self was a spiritual tightwad or emotional anorexic or a redneck with more testosterone than tact (this applies equally to male and female). More often, an artist really is his work, but it's not all of him, or even much of him. Added to that, anyone's brain really is the world, but not all of it, or even much of it.

THE WORLD
IS BREAKING
SOMEONE
ELSE'S HEART

(Emotions)

The Emotional Climate

Who takes
the weeping away now takes delight as well,
which feels
for all the world like honest
work. They've never worked with mind before,
the rich
man says. But moonlight says, *With flesh.*
—Linda Gregerson, "Eyes Like Leeks"

As I was saying, we oversimplify to keep the world stable, consistent, predictable. But we know little of the outside world, only the smattering that leaks in. Mostly we experience ourselves, and roam the mansions of the mind, where emotions can influence so-called facts, a memory tends to gather with its relatives, and one thought teases another. There's plenty of room to swing a cat or a fuzzy emotion. The discrepancy between the actual world and what we choose to know about it is so vast that we often make mistakes, overreact, fill in some of the spaces with superstition. Our ideas may behave, but our emotions are still Pleistocene, and they snarl for attention, they nip at passing ankles. What will it be this morning—love, fear, disgust, sadness, surprise, happiness, anger? Early in our evolution, before the ziggurats of thought, we flinched and gnashed and roared with emotion, wordless, all impulse and appetite. We still do that today, though we hate ourselves for it, for those wild emotions we can only bridle, not break. As Daniel Goleman notes in *Emotional Intelligence,* "The first laws and proclamations of ethics—

the Code of Hammurabi, the Ten Commandments of the Hebrews, the Edicts of the Emperor Ashoka—can be read as attempts to harness, subdue, and domesticate emotional life." All cultures devise laws about how and when to do the harnessing and subduing, and who will do the enforcing. For practical purposes—child rearing, politics, estate planning—we may put emotion and reason in separate boxes and regard them as opposites, even wily opponents. But "they may not be two completely different classes of brain functions at all," as Patricia Churchland points out. "In neural reality, the two are probably part of a continuum."

Not only does our brain do much of its developing outside the womb, we found a way to evolve outside our body, by creating technologies that extend our natural senses. Unfortunately, our brain hasn't caught up with that lurch forward. It doesn't do delicate, subtle emotions graded for each situation. It can worry equally hard about missing a bus, aging, making money, losing a friend. Not in the same way about each thing, but with equal strength. There's not much of a dimmer switch on our emotions. There needs to be, now that some problems call for watchful concern but not high adrenaline, and some rejections won't result in being cast out of the tribe and left to pursue food and shelter all alone. In many ways, we must apply our crude brains to a sophisticated world, and we tend to use emotional hammers when what we need are surgical tools. Life is indeed the best teacher, but the tuition is high.

We evolved to feel anger in familiar arenas, where we could act to make changes or defend ourselves. What we didn't evolve resources for was long-distance anger, fury at potential danger half a world away, and at a level of such complexity and sheer size one can't resolve it single-handedly or even with the help of one's kin. We can feel the requisite anger, we just can't discharge it in useful ways. It's both our privilege and peril to have the brain our hunter-gatherer-scavenger ancestors did, one suited to their equally emotional but simpler world.

When our ancestors became fearful, adrenaline surged through their flesh, and they either ran away or fought. Through both strategies, they survived. Physical threats tended to be life-threatening, and social threats stayed within a small tribe. Today, armed only with the same set of responses, we tackle chaotic and bewildering fears—failing a math test, receiving a bad review, investing badly in the stock market, being in a plane crash, disappointing the boss. For most people, emotions aren't unleashed in mortal struggles, but between family members, neighbors, co-workers. We still surge with adrenaline, but we can't use it as we were designed to, we choke on it. Over time, we created predicaments and habitats faster than our brain could adapt to them. That's still going on today. Our brain is much better at changing the world than living with those changes. The stresses we've invented don't fit our evolution. We're trying to shove square brains through an oval world, and it shouldn't surprise us when it hurts or we can't go on. Many of our valued relationships are faceless, with people we will never meet—corporate executives, government employees, Internet societies. Every single day I receive unwanted e-mail from a company trying to sell me a university diploma. The company disguises the return address and uses a crafty subject line, such as "Question about a student." Or: "Responding to your inquiry." I've tried everything I can think of to make them stop, and spamming me is no use to them since I already have degrees. Yet they continue to send deceptive e-mails. Sometimes pornographic spam will offer me photos of young women with two- or four-legged sex partners. Yesterday I received an ad for a penis-lengthening drug. I find the spammers' tactics irritating, to say the least, and this annoyance sometimes makes me angry. Dealing with it isn't in the purview of the brain we inherited. The best I can do is reach into the brain's basic bag of tricks and pull out anger. Anger spikes the pulse, starts adrenaline pounding, and sends blood to the hands. My own comfort with anger is probably less than most people's. I find anger so disquieting that even when I feel it, my brain likes to pretend I don't. Everyone has favorite emotions and at least one they'll do nearly anything not to feel. For me that's anger. I don't like the feel of its hormones, its sprinting pulse, its tunnel vision—and that doesn't begin to explore the complex psychological resistances I may have to it. In the global scheme of things, a small homely distress doesn't compare with a global anger-monger like war. But it's the small, vicious sprees of fear and anger people routinely face and find more disquieting.

Emotions often provide a dark italics to our lives. Because they can lead to growth, we don't always avoid them. "Afterwards I learned," George MacDonald writes in *Phantastes*, "that the best way to manage some kinds of painful thoughts, is to dare them to do their worst, to let them lie and gnaw at your heart till they are tired, and you find you still have a residue of life they cannot kill."

But, most often, emotions steer us faster than conscious thought. Let's say you swerve on an icy road to avoid hitting another car. In that moment, action blurs: you notice the rush of people, landscape, cars, the shift in gravity, the hoot of a car horn. Those stimuli trigger neurons and your heart pounds, you start to breathe faster, and you recall a similar near miss from years before—a memory flagged by strong emotion. *Beware!* your body says. *Quick: What did you do then?* All of these events combine in a flash to produce that feeling of reverse thunder we call *fear*. Fear can be fast and pure as a traffic accident, but it can also grow as slowly as salt crystals in a cave.

Some years ago I had a climbing accident on a remote Japanese island, where I'd trekked with two ornithologists to see the last surviving short-tailed albatrosses, exquisite white birds with six-foot wingspans, yellow caps, and blue beak tips. The island was an active volcano, camp a small abandoned military blockhouse up ten stories of meandering stone steps. The fishing boat that dropped us off wouldn't be returning for days. And the hundred or so birds nested only at the far edge of the island, in a sloping bowl protected by a plunge of jagged rock. They'd survived the greed of feather hunters because, essentially, they lived in a fortress of stone. My two male companions had upper body strength as well as years of climbing experience, and they could manage a free rope descent— passing a rope behind the back and lowering oneself using brute strength. Unfortunately, I wasn't warned to prepare for those cliffs, and no one brought climbing harnesses. I made it the first day only by straining my muscles. There was no time to question why the men hadn't warned me to prepare—take climbing lessons, lift weights, bring technical equipment. They must have believed I knew the climbing arts, or could improvise them if need be. Why would they have formed that impression? Did I communicate a strength of body and purpose I didn't possess? If so, a deadly bravado. None of this entered my thoughts at the time, when foremost among my brain's sudden scramble was how to survive.

Sleep rested me enough for hiking back across the steaming flanks of the volcano on the second day, but my muscles quivered from overuse. As I jammed a foot into awkward crevices and stretched down, I searched by toe-feel for secure footrests, brailling my way along. When wrist and hand muscles also began to quiver, I ordered my whole body to hold fast, willed it to serve. I imagined both the fast- and slow-twitch muscles (the white and dark meat) tensing elastically, and kept trying, having little alternative. But it was no use; I could feel the electric in each muscle blowing out like a fuse. I couldn't browbeat an untrained body. Physical training doesn't just strengthen the target muscles, it teaches you when you *don't* need to use them, when know-how can replace exertion. I had no muscle memory for technical rock climbing. And I couldn't lower myself parallel to the cliff using only a rope passed behind my back and held taut between rigid arms. Or maybe I could have, had I gauged the right balance and best angle of body parts. Fear began running a slow maze in my brain. Ahead of me, the two men moved fluent as rock spiders, planting and replanting their limbs to grab at a rock edge, shift their balance, or wedge a foot for support.

At last my muscles failed and I fell on the four-hundred-foot cliffs. Fortunately, I held on to the rope or I would surely have died. When I slammed against sharp rock, pain's siren began wailing in my chest, and I seemed to wake up a second time that day. This time to a hard smack in the face, and an order straight from my amygdala. Fear leapt out of its maze and stared me in the eyes. Adrenaline and noradrenaline flooded my system. Heart pounded, blood pressure soared. From spleen to saliva, my body was preparing for further danger, pumping fuel to the muscles, mustering white blood cells, dilating the pupils, shutting down the stomach and other inessential organs, giving the hippocampus something to remember. But I couldn't fight or run. Time stopped. Now what? Crumpled on a ledge, paralyzed by pain, and strangely, I thought, unable to inflate my chest to call out, I said slowly, painfully, crawling along each word, "I'm hurt."

Somehow, with the men supporting me over the last few footholds, I descended to the albatrosses. Because I couldn't raise or lower my body by myself, they dug me a pit in the black volcanic sand, and there I sat watching the fantasia of birds, studying their culture and taking notes for hours. I wasn't numb, but acutely aware of the straitjacket of pain limiting every move. I couldn't inflate my chest enough to breathe fully, laugh, or talk above a forced whisper. Even animating my voice was difficult, because we play the lungs like bagpipes to fill sound with emotion. Yes, I thought grimly, this was the worst pain I'd ever felt or thought possible, and certainly the worst injury. How bad was it exactly?

And yet, a second layer of being settled over worry and anguish like a luminous mental fabric. A surge of rapture and privilege filled my bones, as I watched the courtship dramas of the rare birds I'd traveled half a world to witness and chronicle. (I describe the birds and scene in lengthy detail in *The Rarest of the Rare*, where I barely mention the accident.) They had entered my imagination long before, flown in my dreams and daydreams, focused my energy and resolve, obsessed me royally, and detoured my plans for many months. They'd joined my life at the level of neuron and transmitter chemical, the way a lover or baby can, and even though I'd never met them, they nested in my imagination so completely that finally being in their presence felt ecstatic. For whole minutes at a time, pain, worry, and the outside world didn't exist. Once terror had broken beneath the mind's imaginings, and survival seemed possible, then probable, I felt joy along with the pain. It didn't blot it out. They occurred side by side like the separate notes in a musical chord. You'd think the fear of death would break that spell, but in my journal I wrote, gratefully: "Who would drink from the cup when they can drink from the source?" I had entered nature at the level of feather and stone, my skin stretched to include the complete vista, and in that widening compass nothing lacked. My body had failed me, but my imagination held tight to the rock face it had begun climbing long before, the one Gerard Manley Hopkins pictured in earthy detail when he wrote, "O the mind, mind has mountains; cliffs of fall / Frightful, sheer, no-man-fathomed." My working memory ran its tasks. I remembered how to form sentences and the amount of pressure needed for holding a pen against paper while ink flowed. I threw my eyes and ears at the birds and observed reverently. In my notebook I wrote: "There is a way of beholding which is a form of prayer." I ransacked memory for similes and metaphors to help anchor the sensations. Working memory worked, senses sensed, and I could float creativity like a lit candle over the pond of my awareness. But on another level, my brain was holding an important board meeting of the psyche, fearing death, frantic not to become a permanent part of the all I often craved joining *temporarily*.

Day's end meant being hoisted up the cliffs by rope, and somehow staggering back to camp, where I couldn't lie down or stand up by myself, go to the toilet without help, or roll over. Sleep flattened me. Soon after dawn the following day, still immobilized but hoping I'd improve, I stayed in camp while the two men crossed the island again to view the albatrosses. Lying for hours on the dirt floor in that small cement barracks, I felt sweat welling from my pores, not just on my face but all over my body, which began to shiver. Before then, I'd been fearfully worried but hopeful. Now I became alarmed. Automatically, I began searching the pockets of my cargo pants and long-sleeved shirt for a thermometer. In the back of my mind, that is. In the forefront, a white fog started rolling in, settling in the creases, obscuring the noetic fences.

No thermometer in any pocket. Where was my green knapsack? Pain imposed lightning borders, and I could only reach in a small radius from waist to head. Groping slowly, I found it, the familiar stiff canvas. I dragged it near with one hand and began searching by feel—the fingers outlining each object for my brain to decode. At last I found a slippery Ziploc bag containing small objects, and my fingers and memory told my conscious awareness: *medical bag*. I always traveled with two runs of antibiotics, plenty of aspirin and Tylenol, and other medicines more precious than their weight was a burden. I'd never needed them before. First I removed the thermometer, whose mercury pole I could barely see in the dim light. Under the tongue it went. After five minutes, it read 102 degrees. I took a massive dose of penicillin and two aspirins. Fear goaded my heart into a fast trot, which hurt more with each breath. An hour later, I took my temperature again: 103. By now, I felt cold pouring around me from something like an ice jam, a waterfall. I drank water from a plastic bottle, took two more aspirins together with two Tylenol. Gradually a truth began seeping into my awareness and snagging. *I may die here,* I thought, *on my back in shadows, alone, far from home.* My eyes clung to the dust motes suspended in air at the lit doorway, watched them form small constellations and drift apart. Beyond them the world poured with sun and birds. Each passing hour I took my temperature, which crept higher. I was in a meat locker. Soon I would be as cold as the ground, then the cement. Surrounded by boxes of food and drink, I felt like an Egyptian entombed with supplies for the afterlife.

A noise outside. Animal? Human? Trying to call out, I again discovered I couldn't inflate my chest enough to shout. Pain was a tight suit of armor. A backlit figure appeared in the doorway. Was it real? Hallucination? What was the Japanese word for *help*? I lifted a hand, motioned, pointed to my chest, rolled my closed fists in opposite directions to pantomime the breaking of a stick. A tablet and pen lay beside me. On it I wrote a note to my two companions. I offered the note into dead air, where it disappeared. Then I said one Japanese word very slowly and distinctly: "Ahodori." *Albatross.* There was only one place on the island where the albatrosses nested. The figure evaporated. A person would need to cross the volcano, steaming underfoot in places, and climb down the cliff. No, he might not need to climb. He could yell down to the men. I lay back exhausted. Had someone really appeared? If so, who? From where? We were many kilometers south of the last inhabited island. Would he find the men? And if he did? What could they possibly do to save me? There was no way off the island. Our boat wasn't due back for days, and the medicines weren't working.

Cold became a heavy ingot inside me. When I opened my eyes I saw an angel fluttering in the doorway, haloed in light. She metamorphosed into a beautiful white albatross with long wings and a yellow head. Again she became the angel, floating leglessly into the barracks, then she vanished in darkness, reappearing beside me. Just as suddenly she disappeared. Time passed like a well I had fallen into without hitting bottom. Then the angel appeared again at the doorway, outlined in spangly yellow light. Floating toward me, she knelt once more. But now she removed her wings one at a time and began wrapping them around me. I felt their burning but not their weight. An icy-hot veil encircled my head. I could feel the angel's feathery presence moving slowly, shimmering around in darkness, though I couldn't see her features.

A form passed across the doorway, partly eclipsing the sunlight. I cocked my head like a puzzled dog, trying to make sense of what looked like a newborn baby, fresh from the birth canal, hair matted with sweat. Then I heard birds jabbering in two dialects. Not an angel, but a young Japanese woman sat beside me. Something was on my chest—a long-sleeved white blouse packed with ice. Her profile began to take shape as she turned to speak with the figure in the doorway. I lost consciousness. Time collapsed into itself like one of those portable drinking cups, and I had my arms wrapped around the shoulders of two people as we slowly climbed down the rock stairs to the ocean below.

Did I imagine the angel in the cave who had calmed me with her cool wings until my fever broke? By chance, a charter fishing boat had stopped at the island for lunch, and a crew member had decided she would climb up for the view. Finding me, she felt how hot I was and rushed back down the ten stories, climbing up again with a load of ice from the boat. She took off her white shirt, filled its sleeves and body with ice and packed it around my chest and head. That brought my fever down enough that I could get to her boat, where I was slid into a coffinlike capsule bunk recessed into a wall for the overnight voyage to the boat's home port, the nearest island with a hospital.

There I stared at X rays of my illuminated ribs. Three were

broken. With halves lying horizontally above or below each other, they looked like glowing kindling. None of the organs were pierced, thank heavens. The high fever wasn't from pneumonia but extreme inflammation. The doctor made an open-palmed circle in front of my chest, and said: "OK inside." I pantomimed: *Wrap my chest?* "No," he said, shaking his head, adding a shrug with open hands that I knew meant *There's nothing medicine can do about broken ribs.* He gave me a packet of codeine and cautioned: "Strong . . . maybe night only."

Within a day, I lay on a stretcher-like seat in a jet flying back to the States. The pain still felt torrential, a monsoon prowling my body. To distance myself from it, I replayed a mental film of albatrosses coasting effortlessly on cushions of air. From my memory archives, I took down the volume on albatross courtship, complete with square-dancing curtsies and heads thrown to the heavens in rapturous song. I replayed the expedition slowly, frame by frame. Physical reality disappeared in the magic lantern, periodically returning.

In that cement barracks, with death close and real, my hot brain had cooked up angels and albatrosses. We usually know who's speaking, whether it's us or someone else. But sometimes the body loses track and instead of sensing the outside world and relaying that information to the brain, it generates phantom sensations in the brain itself. When schizophrenics hear voices, they're really hearing, but not with their ears. Instead speech is produced in one part of the brain and received in another. The monitor who decides whom the sounds belong to is off duty.

Deprive the brain of stimuli and hallucinations can flourish. No wonder ghosts tend to appear at night, and spooky houses are usually dimly lit. Blur the edges of reality a little, stop paying ordinary attention, visually recede from your surroundings, and the deprived brain starts frothing with sights and sounds. It doesn't have any choice really, since it abhors a vacuum, and has been keeping tabs on the world since its birth. Anyway, it bores easily and is good at amusing itself. Dim the world, and it generates its own mind theater. Post-traumatic stress disorder flashbacks are different. They're living memories, laid down with the amygdala's blessing, that continue to terrify because they seem real as newly arriving shrapnel.

When I returned to America, I sent the woman a white shirt and my immeasurable gratitude. By return post, she sent me sheets of colorfully patterned origami paper, which were tucked into the envelope like small flags of many nations. I didn't tell her how my terrified brain had refashioned her with wings.

Familiar, comforting imagery can be gripped by the mind, used as a railing to rest on. Extreme feelings that stay abstract frighten more, because they lead to a dizzying lack of control. E. M. Forster's "Only connect" takes on new meaning at such times, when connecting with the familiar matters most. For the likes of me, albatrosses *were* familiars. I'd just spent half a year obsessively studying them, and intense days observing their rituals firsthand. Angels less so. But it was nearly Christmas, and angels were appearing everywhere in the West, an annual migration as predictable as that of Capistrano's swallows. When I first saw the Japanese woman who saved my life, she appeared as one more of those festive angels, who then metamorphosed into an albatross. Both have feathery white wings. She was wearing a long-sleeved white blouse. Flying from fever, my brain seemed at times to be stationary while the world around me flew. I knew I might be fatally injured. The angel, the albatross—both calmed me. This is one example of imagination as a jetty stretching out into chaos, offering planks to the shore. I'm not thinking of chaotic waves in the sense, say, of schizophrenia, when to stop hearing the devil speaking from the drapes one might try to focus on the reality of one's chair or shoe. I mean when the mind accepts the terror of imminent death, with its paralyzing loss of the familiar.

Then fear rouses the senses, and imagination tends to do two things, sometimes alternately: magnify and elaborate the fear until it becomes hugely frightening; or mine the arousal and focus on certain sensations while other information is filtered out. One avenue of philosophy, *phenomenology*, uses this natural skill to tease a thing away from its background, allowing it to float luminously in mental space. Of course, the fearful body does this faster than purposeful thought, and often without conscious awareness, but not always. Sometimes one becomes starkly aware of sensory details, for example the sun sizzle on the waves beside a nearly capsizing sailboat. The mind pays *attention*, a costly currency. In this scenario, terror is temporarily relieved by attending to a sensation, gripping it for a moment, however brief, while the rest of the body goes about its urgent chores. Also, in times of danger, every detail counts because it might provide a key to salvation. Frisking the event with all of one's senses renews the overly familiar, long relegated to the attic of conscious awareness, and dusts it off, rediscovers it, looks at it in new, potentially lifesaving ways.

When does fear escalate to terror? When the element of unpredictability is added? When you realize you may die? Can one be terrified by something like loss of face? Or only by random unpredictable violence or danger? How do childhood terrors (such as the visceral fear of being separated from a parent) differ from adult terrors? Or do they? What role does escalation play? I mean, would what terrified us before terrify us now?

Even simulated horrors frighten us. Like many people, I cover my eyes during the most violent parts of a movie. Not run-of-the-mill horror films, in which killers mainly seem to punish single women for living alone and having jobs. I mean the sort of movie that exceeds cartoonlike violence and enters a zone of realistic, plausible horror. I'm a filmmaker's patsy. I understand the concept of pretend, of let's scare ourselves. I know the violence isn't real, that it was probably concocted on a fake set in Hollywood, and yet I'm unable to use that knowledge as a padlock on my imagination. Counterfeit terror picks the lock and terrifies. I whisper, "Let me know when I can look." In the case of the torture-laden film made from an Aldous Huxley novel, *The Devils of Loudon*, my hand stayed up for most of the movie. I recently watched Sam Peckinpah's *Straw Dogs*, which, when it appeared in the 1970s, scared the pants off many people and sickened others, because it gushed with raw violence. I now find the film's slow-motion violence almost decorous compared with today's film, television, and newspaper violence, depicted in fast, razory detail. Horror, like any other drug, can become the body's familiar and no longer startle. What used to burn only warms. We've gradually raised the threshold, and now require monumental acts of savagery to produce the desired flutter.

It's an entirely different experience to read the war poems of, say, Wilfred Owen and imagine the hell he witnessed. Such stylized secondhand terror, partially digested and ruminated on by someone trying to make sense of it (even if discovering in the end only how senseless it is) is buffered several times. In the process, it becomes language's pet, a growling terror on a leash. Something dilutable, redefinable, prey to persuasion. And that's just enough method to muzzle it. Indeed, muzzling helps convert it from visceral memory to verbal memory, from felt to ignorable.

Sometimes imagination reworks or redefines horror by explaining it as necessary and bearable. In Bible stories, for example, when violence leads to purification, salvation, canonization, resurrection (and countless other shuns). Or patriotically, in wartime. Consider *The Iliad*, a book steaming with carnage, including an unforgettable scene in which a soldier has a spear thrust through his jaw. Not as poetic or magical as *The Odyssey, The Iliad* is essentially an adventure story for teaching young men how to behave appropriately in war. They'd better be able to face its reputed terrors, even desire them. Having heard about them in *The Iliad,* they won't ever forget them. Memory will have stored them as heroic feats. Whatever they find themselves facing will fit into that genre of war horror that others survived with dignity.

For many of us boomers, childhood contained a terror that met us each day like a snot-nosed bully. Who can forget the bomb drills of elementary school, when we huddled under desks and tables, which we wanted to believe would protect us from nuclear annihilation? Whatever *that* was, it terrified our parents more than anything, and became scarier still because of the way it was depicted, as a blindingly bright icon of death. Nothing with details. It was a poster terror—faceless, unpredictable, and no less frightening for being abstract. Is an imaginable terror scarier than an unimaginable one? Not by a long shot.

And yet, those days seem innocent by comparison with today's violence, which we witness first-screen if not firsthand, as extended members of a dysfunctional global family. We're exposed to the raw sights and sounds of horror, which pierce our memories and stick. No abstract icons to terrorize with images of a quick explosive death. Modern terror has teeth, it bites and mauls before it kills. We have a complete inventory of horrible symptoms to imagine, masochistically (or, if you like, superstitiously), and imagination likes rich grit to pearl around.

We live in an image-mad era, but for most of human existence words ruled, and words take time to analyze. They're too abstract for the brain to swallow whole; they need to be reimagined in the mind's eye. Images, on the other hand, are like visual punches. They're in your face, with no time to duck and no way to imagine things differently. Prepackaged horror, salted to someone else's taste, can make you gag. Nor are the images the poor facsimiles of long ago. They feel real enough to blast our emotions and create new memory circuits. When you see the images again, the circuits grow stronger. Some sights may ache long afterward.

Deluged with media violence, our wits obducted, we're forced to be helpless viewers of other people's pain. This morning, television news brought unbearable suffering into my living room. Shouldn't I rush to California to comfort wildfire victims? Shouldn't I fly to the Middle East to aid bomb victims? On another channel, a show dished up real (I think) cops and real (I think) sexual offenders. Shouldn't I be afraid, too? No wonder the limbic system cringes and bolts. A PET scan would show the addled bloodflow in my brain, as I felt sympathetic fright, dread, and anger. An unnameable sin seemed to weigh me down. There's no remedy for such colossal guilt except to tune it out, and we do, as we gradually become numb to the suffering of others.

Over a thousand studies suggest that media violence inspires real violence. The younger the child, the greater the influence, since children have more trouble telling real from unreal, here from there, past from present, sometimes from always. But adults are vulnerable too, especially those already anxious or depressed. Images of fear, horror, danger, and violence hit the right hemisphere harder than the left, and people already dominant in that hemisphere may find themselves feeling so stricken and overwhelmed that even the left hemisphere's storyteller can't calm them down.

Because we've been bombed at home, in our neighborhood, we've become prey keeping watch for glowing eyes stalking us with lynxlike hunger. This reminds us of something we'd rather not acknowledge, that we're only at the top of the food chain by chance, not because we're faster, stronger, or better armored than other animals. Nor because we have a larger biomass or longer lineage. What we are is mindier. An important part of such terror stems from our biological fear system, a hair-trigger one that guided us before we became what we like to refer to as civilized. In our ancient past, we were always on the lookout for an unexpected flex in the grass or the sound of claws scratching on rock. Our physiology is still primed to respond quickly and extravagantly to even the possibility of harm. And to crave its opposite.

CHAPTER 26

The Pursuit of Happiness

[Happiness] is really something effervescent that fills
me completely with a light, pleasant quiver and that per-
suades me of the existence of abilities of whose nonexis-
tence I can convince myself with complete certainty at
any moment, even now.
                                        —Franz Kafka, *Diaries 1910*

We know so little about happiness, that ultimate oasis we all
hope to find and never leave. It's not epidemic, doesn't
alter the balance of power, won't melt pounds or sap wealth. So it
doesn't make the news much or fetch many research dollars. Ironically, although its presence is the all we wish for our loved ones, we
mainly study its absence.

There's petit mal happiness, when you're beguiled by such nov-
elties as overdue praise, dry heat, ripe apricots, plenty of anything,
great sex, or being chosen. And there's the grand mal happiness of
sprawling stickily in love. Happiness can occur when one isn't
looking. "In retrospect," one might say, "those were my happiest
days." Were they really happy? Or is memory sugaring them? Was
Proust right: Is the only paradise a lost or remembered one? Was
Pascal right: Are we most happy when dreaming about future
happiness? Can there be a torrent of happiness, or only moments
of being, as we compare the luminous present with the fading millisecond before? A sweet calamity of pleasure can float like a tropic
island in an otherwise humdrum day. Sometimes minute flakes of
happiness link up, and the doubtful brain quizzes itself: *Am I still*

happy? Yes, I'm still happy. Better check. Am I still happy? I think so. Hold on, now I'm not as thoroughly happy as a shred of a second before. Okay, I'm happy again. And so on. These jaunty little tricks of mood become invisible beads on the single strand of a day.

Our Constitution doesn't guarantee us the right to possess happiness, only to *pursue* it, and we pursue it hotly, though not always safely. That we believe happiness *must* be pursued says a lot about us and how ephemeral we find it. One only pursues something that's out of reach and escaping—a stag, a nymph, a star, a love. We pursue happiness like the wild and dangerous creature it is. Why *dangerous?* Because it can turn on you unexpectedly and become its opposite. Because its absence can be more powerful than its presence, and it's always present in its absence. Because it promotes contentment, which leads to inertia, even though some people are happiest when gyrating physically or mentally. Most people feel happiest when still, not hushed but tranquil, and not *un* anything—untroubled, unagitated—but full of a lucid calm that can't abide. They pursue happiness as a fixative, an ultimate goal, after which they can coast downhill. Dangerous because happiness tends to elude by inches, as the poet Robert Browning understood when he wrote: "a man's reach should exceed his grasp, / Or what's a heaven for?" And because happiness has a tendency to sour after a while. It's perishable. When well-being becomes familiar, we take it for granted, and from there it's an easy crawl to boredom and less happiness. "Not another parachute jumper," you sigh after watching twenty of them on a sweltering August day at the Oshkosh Air Show in Wisconsin. "There is no doubt that happiness makes one idle and sterile, and that perfection can become very boring, and that an existence in Dante's *Paradiso* would be unbearable to us," Sigmund Freud wrote in a letter to his patient, the poet Hilda Doolittle, on October 5, 1933. "And yet we wish each other happiness and contentment and other inappropriate things."

Is happiness an addition or a loss? Is it a comparative state, a response to a change for the better? When the body regains its homeostasis, does the brain reward it with happiness, a gold star for a job well done? "Happiness is equilibrium," the playwright Tom

Stoppard offers, "Shift your weight.... You compensate, rebalance yourself so that you maintain your angle to the world. When the world shifts, you shift."

We have the peculiar fate to be living beings made up of non-living chemicals, and there's the heart of the problem. Both mind and matter, we learn as infants that if we don't mind something, it doesn't matter. If we do mind, a fear can loom to the size of a supercomputer in a sci-fi story I once read, a gleaming hunk of computer circuitry that took decades to build, and which eager scientists switched on for the first time and asked: "Is there a God?" A booming voice answered: "There is now."

The mind has its broom closets, where it stores contraptions that aid in forgetting. Forgetting in order to be happy. Otherwise painful flotsam from the past would pollute the sunny atoll of happiness, where one can only thrive if one ignores the shadows in the shallows, where the sharks of misgiving lurk with endless appetite. We have a bias for happy memories. Because all memories were once sensations, the bridge between present and past is inevitable, always open, with traffic jostling in both directions. Sometimes we motor sadly to where we dread arriving: perhaps a cemetery where a few months ago mother was buried beside father, who had died two years earlier. Grief sifts through one's memories, sampling and organizing them in strange ways. When I miss her fiercely, my mind reaches for good memories, good features. Mother teaching me how to take a bath (cleanest parts first), how to shave my legs, how to shape my eyebrows, how to walk in high heels, how to prevent wrinkles (remove makeup with baby oil). For a while, later on, she had *me* teach *her* how to apply makeup and do her hair, and then, finally, how to help preserve her deteriorating health.

I sense these memories, which I don't think so much as feel, just as I feel the childhood excitement of snow falling like a big toy. I wasn't with her when she died, but I believe she spent her last hours in a cocoon of happy memories, lit in part by opiates and other drugs, and emotionalized by the limbic system. Most likely she voyaged back to her childhood in Crisfield, Maryland, a magic time that hung like a lantern in her past, and which she often spoke of with deep nostalgia. Only in the last year of her life did she admit to being old, and then it was to lament that she felt like a young girl trapped in an old woman's body. At eighty-two, despite acute leukemia, she told none of her friends about her illness, because they might hesitate to introduce her to cute guys. She went to singles dances, still tried to look *glamorous* (a word that meant the world to her), kept her passion for travel, and obsessed like a teenager about finding a sweetheart. She often said, wistfully, that she wished my brother and I were little again, how she preferred us as little children, and from the vantage point of old age, I'm sure those seemed golden days, happy despite the hardships, because she was young and beautiful then, and her children were living toys who worshiped and adored her. The hard times she was happy to forget, needed to forget in order to feel happy.

A born optimist, she took credit for her successes, and believed setbacks couldn't stop her. She rarely blamed herself. She decided to have had a happy past. She kept busy and socialized nonstop. She *chose* to be happy, to define herself as an upbeat person who could face any hardship and prevail. She inherited much of her happy disposition from her father. There's no one happiness gene, but humans are predisposed to be at least mildly happy, which bolstered our ancestors, who needed to stay hopeful and active to survive. Twin studies show that as much as 50 percent of one's sense of happiness may be hereditary. A constellation of genes can favor qualities like optimism, openness to new experiences, sociability, and other qualities my mother had, which combine in ways more likely to yield happiness. But, as I said, she also actively pursued it on a daily basis.

As the University of Pennsylvania psychologist Martin Seligman argues, happiness can be learned, just as helplessness can. He's founded an institute for "positive psychology" and the formal study of happiness. *Finding the bright side,* favoring positive explanations, consciously doing things you know will make you happy, playing more, choosing enjoyable tasks, cultivating optimism and hope, pursuing happiness—are among the transformative skills to practice. "Anyone can be sad when they're sad," my mother used to say. "The trick is to be happy when you're sad!" Not easy, and I must confess I often found her edict annoying, because it dismissed the seriousness of my feelings. But for her it worked to "accentuate the positive," as a song of her era advised. So she loved *Cultivating Delight: A Natural History of My Garden,* with its bedrock urge to forget winning or losing, just cultivate delight. I finished that book one summer, while in a collar with a broken neck, anxious about an operation that might leave me paralyzed. Yet each day I rose happy, rubbed my mental hands together, and said: "Oh, boy, what can I learn today?" It was one of the happiest writing experiences of my life, full of nature study, surprise, mystery, and marvel. Wonder is a bulky emotion. When it fills your heart there isn't room for anything else.

A 1994 study of happiness in sixty-four countries found that people who report being happy tend to be married, outgoing, and healthy. Whole countries were happier than others, with Scandinavian countries leading, perhaps because some of the survey's criteria were national peace, economic development, government stability, minimal class consciousness, modernity, and gender equality. Add to that national healthcare and retirement benefits, and many of life's organic stresses are softened. Some people still felt discontent, plunged into depression, succumbed to *morketiden*— deathtime, when there's so little sunlight one's circadian rhythm starts to stumble, and *cabin fever* can't begin to describe the sense of panic. People sometimes drove their snowmobiles toward the icy horizon, hoping to homestead it, until they ran out of gas. But most people enjoyed a strong sense of contentment and well-being.

If I strolled through Scandinavia, would I find more smiles per mile? I mean natural smiles, ones that well up from inside and flood the face with light. We can't control the soft eruption of real smiles. The University of Iowa neuroscientist Antonio Damasio finds that we use different "machinery," different brain circuits, to fake a smile. Spontaneous happiness involves the frontal cortex and the hypothalamus. "So let's say that you damage the left motor cortex," Damasio offers, "and as a result you have paralysis on the right side of your face, and, in fact, the face becomes asymmetric. Then whenever you tried to smile only half of your face would rise." But if you hear a truly funny joke, your face crinkles into a smile. "And at that point, both sides of the face move quite evenly in the kind of natural smile you had before you had the stroke."

Is a happy disposition a genetic gift? At the University of Wisconsin's Brain Imaging Laboratory, Richard Davidson, a professor of psychology and pyschiatry, has been studying people with two kinds of temperament. One group is more vulnerable to setbacks and stresses, more easily stymied, prey to unhappiness. The other group springs back quickly from life's disappointments and catastrophes, is less often depressed, feels generally optimistic and hopeful. Using PET scans and fMRI, Davidson peers into the brains of people watching happy or sad images—a loving mother cosseting her baby, a badly deformed child. Resilient, positive people show more mindglow in the left prefrontal cortex, and an inhibited amygdala. In negative, more vulnerable people, the amygdala springs to life and the right prefrontal cortex becomes lively. At Harvard University, the psychologist Jerome Kagan, who has been studying a group of children for over a decade, also finds that happy kids and adults tend to have a more animated left prefrontal cortex, "while the shy, serious, apparently less happy children tend to show more activity in the right frontal area." Does this mean that a child born with a dominant right brain is doomed to a life of sadness? Maybe it's "born with a bias," Kagan theorizes, "but not a deterministic one . . ." A bias that good parents, friends, school, and environment can ameliorate, though not banish. "It's just that he's got to fight the bias." Someone born with a gene for a rich contralto voice may become a great singer, but someone without the gene may fare equally well, if she's passionate about singing and trains hard. "The brains of small children are still highly malleable and easily influenced by experience," says Kagan. And not just by deprivations in the womb or early infancy. One's happiness can be powerfully influenced lifelong by parents, birth order, ethnic background, world events, role models, personality of siblings, fate of loved ones, illness, trauma, genes, luck, and many other factors.

Happiness doesn't require humor, but they often travel together. Much of what's funny depends on incongruity, something familiar in one context appearing out of place, a topsy-turvy situation. Here are a few of my favorite instances of this sort of humor: President Lyndon Johnson once quipped that when he died, he hoped to be buried in Texas so that he could stay active in politics. After a campaign rant by a bigoted politician, Molly Ivins was asked if she enjoyed the speech, and she replied dryly, "I liked it better in the original German." I think it was also Ivins who said of a politician, "If he were any smarter, you'd have to water him." When I visited the University of Arkansas some years ago, I really enjoyed the local accent. I knew there were famous hot springs in the southern part of the state, so I asked my host: "Are there any spas in this part of Arkansas?" He looked puzzled for a few moments, and finally asked: "Do you mean Russian agents?" In Steve Martin's fictions, he balances intellect with supreme silliness—a combination I relish. In his "Disgruntled Former Lexicographer," a man who has worked on Random House's dictionaries for thirty-two years is fired after the following definition of *mutton* is found in a 1999 edition:

mut-ton (mut'n), n. [Middle English, from Old English *mouton*, *moton*, from Medieval Latin *multo, multon-*, of Celtic origin.] 1. The flesh of a fully grown sheep. 2. A glove with four fingers. 3. Two discharged muons. 4. Seven English tons. 5. One who mutinies. 6. To wear a dog. 7. A fastening device on a mshirt or a mblouse. 8. Fuzzy underwear for ladies. 9. A bacteria-resistant amoeba with an attractive do. 10. To throw a boomerang weakly. 11. Any kind of lump in the pants. (*Slang.*) 12. A hundred mittens. 13. An earthling who has been taken over by an alien. 14. The smallest whole particle in the universe, so small you can hardly see it. 15. A big, nasty cut on the hand. 16. The rantings of a flibbertigibbet. 17. My wife never supported me. 18. It was as though I worked my whole life and it wasn't enough for her. 19. My children think I'm a nerd. 20. In architecture, a bad idea. 21. Define this, you nitwits. 22. To blubber one's finger over the lips while saying "bluh." 23. I would like to take a trip to the seaside, where no one knows me. 24. I would like to be walking along the beach when a beautiful woman passes by. 25. She would stop me and ask me what I did for a living. 26. I would tell her I am a lexicographer. 27. She would say, "Oh, you wild boy." Exactly that, not one word different. 28. Then she would ask me to define our relationship, which at that point would be one minute old. I would demur. But she would say, "Oh, please define this second for me right now." 29. I would look at her and say, "Mutton." 30. She would swoon. Because I would say it with a slight Spanish accent, at which I am very good. 31. I would take her hand and she would notice me feeling her wedding ring. I would ask her whom she is married to. She would say, "A big cheese at Random House." 32. I would take her to my hotel room, and teach her the meaning of love. 33. I would use the American Heritage, out of spite, and read all the definitions. 34. Then I would read from the Random House some of my favorites among those that I worked on: "the" (just try it); "blue" (give it a shot, and don't use the word "nanometer"). 35. I would make love to her according to the O.E.D., sixth definition. 36. We would call room service and order tagliolini without looking it up. 37. I would return her to the beach, and we would say goodbye. 38. Gibberish in e-mail. 39. A reading lamp with a lousy 15-watt bulb, like they have in Europe.

Also: a. muttonchops: slicing sheep meat with the face. b. muttsam: sheep floating in the sea. c. muttonheads: the Random House people.

When I got to definition 6—"To wear a dog"—I began laughing spasmodically, and again at odd moments throughout the day, whenever the image of wearing a live squirming mutt infiltrated my thoughts. At dinner that evening, over curry, I tried sharing its humor with friends. Only one of three got it, and she started laughing crazily, too, while the others seemed mystified by our apparent stomach cramps and bad taste. On the basis of that single event, I conclude that humor is subjective.

If you're like me, you're wondering who told the first joke, and what sort of joke it was, if anyone got it, and when laughter began. Researchers at the University of Hertfordshire, in England, say they've sampled popular jokes in different countries and selected the World's Funniest Joke. It goes like this: A couple of hunters are out in the woods when one of them falls to the ground. He doesn't seem to be breathing; his eyes are rolled back in his head. The other guy whips out his cell phone and calls the emergency services. He gasps to the operator: "My friend is dead! What can I do?" The operator says: "Take it easy. I can help. First, let's make sure he's dead." There is silence, then a shot is heard. The guy's voice comes back on the line. He says: "OK, now what?"

Not a laugh riot, but not bad, and it juggles the right balls of surprise and incongruity to strike us as funny.

Most laughter has little to do with humor, though. "There is a dark side to laughter that we are too quick to overlook," the psychologist Robert Provine reminds us. "The kids at Columbine were laughing as they walked through the school shooting their peers." Laughing has more to do with socializing. Laughing together, we enter synchrony, especially if we laugh with loved ones. Hence the tickle laughter between parents and their children. Chimpanzees are also great laughers, though they do it differently, on the inhale, rather than on the exhale, as we do. They adore being tickled. Why does laughing feel good? Endorphins? Oxygen flush from the hyperventilation? Calming of stress hormones? Whatever its origins, it makes play fun, and play is serious business both as tutor and as cementer of relationships.

What caused the first laughs? Watching a prank? Enjoying the misfortune of someone else? "Humorists fatten on trouble," E. B. White once quipped. Courtship and play fighting? A spontaneous laugh includes baring one's teeth—normally a sign of aggression. Laughing squeezes the eyes small, and one may even sway or become spastic. Facially, a laugh falls somewhere between a grinning smile and a vertiginous yawn. In a fight one fixes eye blades on the enemy, stays rigidly alert. Laughing makes us vulnerable to attack. Maybe what it signals in love or war play is mock aggression, something like: *Don't retaliate by killing me. I'm only playing.* Maybe it evolved to promote intimacy, affection, trust, the secret of a shared mood.

One day when surgeons were operating on a woman's brain, they accidentally touched the orbital prefrontal cortex and were surprised by her sudden laughter. Since then, scientists have found that the region lights up during a joke, especially a short one. The Dartmouth neuroscientist William M. Kelley and his colleagues tested people while they watched an entire episode of the TV sitcom *Seinfeld*. During a funny scene, MRI images showed a surge of activity in left-hemisphere regions associated with resolving ambiguity, says Kelley, who goes on to add that seconds later, both hemispheres showed activity in other regions, associated with emotions and memory. Did laughter emerge as a mental brake, a way to slow the brain down long enough to make sense of illogical situations?

Along these lines, archaeologist Steven Mithen concludes that the Neanderthals and their predecessors didn't have a sense of humor, couldn't really, because their brains weren't able to pull together elements from different domains. Before cognitive blending, incongruity wouldn't have made sense, or nonsense, to them.

I think humor rides many razor-backed pigs, even if illogic and incongruity seem to dominate the herd. I doubt there's just one explanation for laughter, because most of what we laugh about isn't really funny. We laugh when we're embarrassed, nervous, or uncomfortable. We laugh to change the mood of a conversation. We laugh sarcastically. We laugh to ridicule. We laugh in fond greeting and farewell. We laugh when surprised, during flirtation, simply for pleasure. We laugh to forge alliances. Humor, on the other hand, as distinct from laughter, may be something uniquely human. But play is enthusiastic and widespread among mammals. It sharpens the senses, builds muscle strength and coordination, and helps animals rehearse for adult life. (Because I explore the world of play at length in *Deep Play*, I won't spend longer with it here.)

Even rat pups make high squeaky yelps when they play, noises adult rats don't make as often, and some scientists interpret that as the rat version of laughing. Laughter seems to be an old device, shared by many animals, generated long ago by the limbic system. People in all cultures laugh, and most tellingly perhaps, even blind, deaf, and dumb children, who have never seen or heard laughter, laugh naturally. Fun is its own reward, and guides animals well. Angelic and devilish, without a trace of shame, evolution uses fun and pain to train its creatures. Laughter disarms and instructs, but it also feels so good that we'd laugh even if it were taboo. Maybe in the sort of dimly lit laugh dens we call comedy clubs.

Back to the limbic system, where laughter slithers. Does this mean that toads and lizards laugh? I've never met a laughing alligator, although in a children's book I did write about a toothy one with a permanent smile. No doubt my ever-laughing dentist has a very active hypothalamus, which seems to be involved with loud, rolling laughter. When he "gets" a joke, he's most likely using a brain circuit that ripened in humans to aid in communication and the fine control of facial muscles that make both talking and laughing possible. What began in play and tickling, we transformed into jokes—something Darwin regarded as a form of psychological tickling, and Theodore Reik as a shock without alarm. Building on that basic design, we laugh and play with others, with lovers, and with family, in some of our finest moments.

Happiness doesn't require laughter, only well-being and a sense that the world is breaking someone else's heart, not mine. For us, I mean. We tie balloons of thought to our feelings. Animal happiness doesn't require a sense of self. Spanieling is a good example. Here's my definition of that activity, which I regularly enjoy in winter:

> spaniel *v* (fr. *spaniel,* any of several breeds of medium-size dogs): To find a shaft of sunlight pouring through a window on a cold winter day, curl up in the puddle of warmth it creates on the rug, and doze with doglike dereliction. "I think I'll just spaniel for an hour or so before I begin work," she said.

THE COLOR
OF SAYING

(Language)

CHAPTER 27

Memory's Accomplice

Perhaps we are *here* in order to say: house,
bridge, fountain, gate, pitcher, fruit-tree, window
. . . to say them more intensely than the things themselves
ever dreamed . . .

> —Rainer Maria Rilke,
> "The Ninth Elegy"

Babies are born citizens of the world, as linguists like to say. It doesn't matter if they're born into a world of high-rises or tundra, jackhammers or machine guns, Quechua or French. The ultimate immigrants, babies arrive ready to learn the language of their parents, with a brain flexible enough to adapt to almost anything. Whatever language they hear becomes an indelible part of their lives, provides the words they'll use to know and be known. If two languages are spoken at home, they'll become bilingual. One of my nieces is trilingual, because her Brazilian mother spoke Portuguese as well as English to her from birth, and then together they learned Italian. A bonus of bilingualism is that it forces a child to favor one set of rules over another, and that trains the brain early on to focus and discriminate, ignore what's irrelevant, and discover how fickle words can be.

My mother's father spoke seven languages well—Polish, German, Hebrew, Russian, Yiddish, English, and Gypsy (a Romanian dialect). As a child, growing up on a farm in Eastern Europe, he and his siblings traveled around the countryside to sell eggs and produce, and Grandfather was just young enough to learn languages easily. How did he do that? Most of the folk he met couldn't read or write and didn't understand the rules of grammar any more than his own parents did. Yet he quickly absorbed the complicated grammar for six languages while still a child. MRI studies show that when kids learn a second language as toddlers, they use the same part of Broca's area for both. To the brain, it's all one language. But once they're older, learning a new language is hard work that recruits other parts of Broca's area. My grandfather learned English as a young adult when he came to the United States, and he always sounded like he was speaking a foreign tongue.

At around six months old, babies start to identify the special sounds of their native speech, like the umlauted *u* of German that requires a little lip pucker, or the squeaky *e* of American English's *street*. Before their first birthday, they can analyze word order and memorize sentence and sound patterns. Long before words make sense, babies learn a circus of familiar sounds, all the exotic vowels and leaping rhythms. Babies the world over babble alike at first, then gradually babble in their own language. Children born deaf babble with their hands. But we're not the planet's only babblers. Birds and monkeys babble, which suggests that babbling evolved long before language, perhaps as a plea for affection or to summon Mom. In that case, language may have bloomed from a natural urge to babble. One of the brain's best tricks is how fast it sponges up language. In its first few years, the brain is so plastic, so busily composing itself, that it can almost inhale a language.

A kid's job is to be cute so that grown-ups will love and protect it, and cuteness requires childlike features, everything from big eyes and rounded face to naïveté. Language-learning mistakes rank high on the cuteness scale. So high, in fact, that grown-ups, being cute on purpose, sometimes talk baby talk, including mistakes in pronunciation and grammar. In a notorious one-line review of a sappy children's book, Dorothy Parker, also known as Constant Reader, once wrote: "Constant Weader fwew up."

Children learn grammar just by hearing it, without being taught, and then they mistakenly apply its basic rules to everything. The plural rule in English is a good example. We most often add an *s* to nouns to pluralize them. The plural of spouse isn't spice, though, personally, I prefer it. Children learn the rules before the exceptions, and so make plurals by adding *s* to everything (*spoons, cars, mouses, foots, gooses*), sounding quintessentially cute in the process. Cute enough to cuddle, gab with, and teach more language.

How miraculous human language seems. But no more so than hummingbirds being born with the ability to navigate through jungles, over mountains and across seas; or bloodhounds with a talent for discriminating among thousands of odors. Language is our plumage, our claws. Because species evolve what serves them best, the ability to decipher complex rules of language is woven into our genetic suit. Many animals have all the tools and skills needed for speech. We even share an anatomical felicity long thought to be unique to humans—the low larynx that makes speaking manageable—with red and fallow deer. What they don't have is our knack for combining a few letters, sounds, and words in infinite ways. We can do that with speech, just as nature does that with . . . well, most everything. How many ways can you use a sieve? How many forms can an eye take? Science may call that convergent evolution, but it really has to do with elasticity, willingness, and a yen for experiment. What works wins. *Give it a try* is nature's implicit creed and our spirited motto.

Human babies learn language the way most baby birds learn their songs, by imitating grown-ups. Like birds, we have special language areas in the brain and a learning window. A bird or child raised in isolation, then introduced to its song or language later in life, won't be able to fully master it. Learning a language as a grown-up is heavy lifting. A friend's son, who is spending a year as an exchange student in Slovakia, sent home this wonderful account of a smart, determined eighteen-year-old's adventures learning to speak Slovak:

1. Genders—As with many languages, adjectives change according to the gender of what you're talking about. It can be a guy, it can be a girl, it can even be neuter (or as I say, spayed). So there are three different endings to each adjective. As if that wasn't enough, the past tense also

takes on the gender of whatever it is you're talking about as well. Basically, verbs, nouns, and adjectives all have genders in Slovak.

2. Imperfect and perfect verbs—I don't really know how to explain this one very well, because I still have to get the hang of it. Basically it differentiates between a regular action and a onetime occurrence, but it's more complicated than that. Unfortunately, since there are so many types of verbs, it is hard to tell when they are perfect and imperfect. Take the imperfect—to make it perfect you might stick a prefix onto it or you might make the ending longer somehow, or just change the ending slightly, or in fact, it could get shorter, there are patterns, but there are plenty of more important things, like . . .

3. Pronunciation—Not wholly difficult so much as wholly different. Here're some letters: æ, ö, Ë, ù, û, ·, Ì, È, (tm), ", Ù, Ú. Here're some words: "k," "zmrzlina," "kæ(tm)Ë," "prst," "vûdy," "dlh." Are these common words, you ask? Here's what they mean: "to (someone's house)," "ice cream," "key," "finger," "always," "long." Probably every day I use all of those (maybe not *ice cream*). Did I mention that vowels with those marks on top are long vowels? Like, if you've got "a" and "·," they're the same noise, you just say one of them longer. Did I mention soft consonants? Yeah, you've got your *d, t, n, l,* but when they're in front of an *e* or *i* or they have those marky things, you make them a soft consonant, which is like adding a yuh sound on the end—don't try, it's not worth the trouble.

4. Plurals—Depending on the gender and ending letter of a word, its plural is different. Some are totally weird and random, like *ears* and *eyes,* but that's true in English too, person-people and what not. Side note: The plural of *ear* is sort of the equivalent of *earses.* When I asked my host brothers friends why ears had a different plural (WARNING: Do not ask Slovak teenagers about Slovak grammar), they told me that it's like the word *pants* in English. Then I said, "No it's not, because *ear* can be singular too" and they told me, "It's like pants." I said, "You aren't listening," explained again, and

they told me, "It's like pants." I hate that sentence. It's not like pants.

5. A thousand words for everything. Some argue it's the same in English, but I don't agree, I am right, everyone else is wrong. That's all there is to it. When you have a different word for cutting with scissors and cutting with a knife, it's gone too far.

6. CASES—This is the big mama. After ANY preposition (with, to, in, about, onto, from, etc.), the end of a noun changes. Let's take the word *auto* (car). Its forms look like this: auta, aute, autom, autoch, autov, autom, autami. And the plural aut·. That's why it's tough. Also, if it's the subject of a verb and not the acting out of the verb, there is a different case as well, so it's possible that there is a case without any preposition whatsoever. I feel like I missed a reason. Oh well, I am sure that's enough for that.

Anyways, regardless of all of that, people tell me that I am learning quite quickly. I didn't think I had learned all that much until the rotary weekend in my town when we were all sitting in our big van, I was in the back, and I remembered that two students had written me to say they were coming late so I told the coordinator in Slovak and the van went quiet until the American girl sitting next to me turned to me and said, "I hate you." Honestly though, it's not even my fault that I know Slovak as well as I do, I am just really determined to do things well, it would take more effort to force myself to not learn Slovak. I think about it wherever I go and I always ask people questions . . .

His account makes the case beautifully, I think, for the ease of language learning as a child, and the tribulations that wait beyond the magic window. Language is so hard only children can master it.

We use words to label and categorize, to discern subtle differences, to group related things, to build endless lists. But also to create false divisions, false distinctions, false unities, which become possible at the moment they're put into words. For example, in one of his nonfictions, Jorge Luis Borges refers to a "certain Chinese encyclopedia" in which animals are divided into

(a) those that belong to the emperor; (b) embalmed ones; (c) those that are trained; (d) suckling pigs; (e) mermaids; (f) fabulous ones; (g) stray dogs; (h) those that are included in this classification; (i) those that tremble as if they were mad; (j) innumerable ones; (k) those drawn with a very fine camel's-hair brush; (l) et cetera; (m) those that have just broken the flower vase; (n) those that at a distance resemble flies.

It's not that we can't imagine a category that only includes animals "that have just broken the flower vase." We can, because we've classified things in equally absurd ways at times. Browsing through the dictionary one day, I came upon this extremely esoteric category: "hapax legomenon," which refers to words that occur only once in the entire body of a dead language. One day, soon afterward, when my sweetheart asked me what planet I came from and if there were any more at home like me, I answered smugly: "No, I'm the hapax legomenon." Allowing a whole category to stand for an individual is something humans love to do with words. The opposite is synecdoche—when part of something stands for its entirety. Most curse words fall into this category. There are public associations and private associations. A friend of mine once told me that she was "waiting for the pecs to arrive," by which she meant the muscular college football players she'd hired to spread manure on her flower beds. For her purposes, on that day at least, they *were* their chest muscles.

It's amazing how much information the ever-simplifying brain thinks it worthwhile to store in words. Do I need to know that a typical acre of land contains about fifty thousand spiders? Maybe so. I haven't experienced them all personally, but I can generalize to the leggy mob, and hope they're hungry enough to eat the aphids plaguing my roses. Do I need to know the only four words in the English language that end in -*dous* (hazardous, tremendous, horrendous, stupendous)? Probably not. But my brain's yen to make categories takes note and finds it curious.

Can we think without words? Sure. I paused half a dozen times in the last page to find the right word for something I was thinking. I tried out several words or phrases each time, and also reconfigured a couple of sentences. At the University of Nevada, Russ Hurlburt gave students beepers and told them to jot down whatever was in mind the split second their beepers went off. The surprising result was that people talk to themselves only 32 percent of the time. For 25 percent of the time their mind is full of imagery, for another 25 percent they have a clear sense of thought but it's nonverbal, and for the rest of the time they're feeling pain or some other emotion. Notice, for much of our lives, or those of University of Nevada students anyway, the mind is humming along wordlessly. Forcing the students to jot down their thoughts at odd moments turned up all sorts of unexpected preoccupations the volunteers didn't know they were obsessing about (because they normally chose to forget their anger at so-and-so or worry about whatzit). Most of the time we don't think about what we're thinking about, and we're thinking wordlessly. When we start to monitor our ghostly thoughts, they assume form and become visible, usually by donning words.

Thanks to language, we have a verbal memory that allows us to learn and remember without physically experiencing something. Pure magic. Thanks to writing and its kin, we no longer have to memorize the endless fine rubble that passes for everyday life. We make lists, we take notes, we file things away. Books invite one to view another's mind, self, suite of defining memories. Instead of straining to remember everything, we can deploy our attention (and many neurons and synapses) to toil at other jobs—coining new games and ideas, for instance. Most often, we don't know what we're feeling or thinking until we put it into words. Words can act like tongs to drag a squirming idea into view, or like a set of lenses to help focus our thoughts. Words can cap gushing emotions, and trawl for memories. They can highlight and frame things when we need perspective, and they're excellent handles when we need to grip a slippery notion. As social beasts, we trade words with others, negotiate meanings, use words as currency. Words form the backbone of what we think. So, although it's possible to have thought without words, it's rarely possible to know what we think without bronzing it in words. Otherwise, the thoughts seem to float away. Refine the words and you refine the thought, but that often means squishing a square thought into a round hole, and saying what you can instead of what you mean. We try to remedy that by piling up words like brushstrokes in what we call descriptions, or explanations, or by blending images ("words," "paint," "brushstrokes"), or by adding emotional sounds to what we say. "Please do that for me" means altogether different things if you say it pleadingly or in separate jabs. Yesterday I recommended *Analyzing Freud: H.D.'s Letters* to a friend I thought would relish the book, and he thanked me for the suggestion with the slightest edge in his voice. We don't have a word for that often-used sound that means: *Oh, no, not something else to do, I'm already up to my eyeballs in work and the last thing I need is another book to add to the stack, no matter how much I'd truly enjoy it, but it does sound wonderful, and I don't want to refuse your thoughtfulness.* All of that can be conveyed with tone.

Using words can also sabotage understanding. For example, *punishment* and *reward* are two overly-calm-sounding words for a vast, seething array of lures and deterrents. Language makes it possible to say those words without tasting their reality in memory excursions. We *can* remember instances of both, if we want to, or are driven to, but we don't have to. Instead, we may store them on a mental shelf like canisters, tightly lidded, their contents vague and hidden. This natural gift leads to another sleight of mind. Eyewitnesses to a crime who write down a description of a robber are much more likely to make memory mistakes about him later. They can observe in richer detail than they have words for. When they try to recall his face, they remember what they wrote rather than what they saw. Language blots out observation in what's known as verbal overshadowing. Because life eludes words, events can't be fully recorded. Yet words do serve as nimble memory aids for some people. For them, finding the right words, words accurate enough to trigger a robust sensory memory of what sponsored the words, becomes a holy quest.

I'm perpetually amazed by how eager humans are to complicate things. Isn't language complicated enough? Apparently not. Every family invents its own special dialect, a sort of tribal speak, as members bring home this or that slang from school or work, and add televisionese or song lyrics to the general mix. A separate lingo binds people, but I find another motive persuasive, too: our endless need to express the sheer feel of being alive. How does the brain convey that to itself and others? Only by using as many tricks as possible, preferring one in particular.

CHAPTER 28

Metaphors Be with You

"Madam, if I were Herod in the middle of the massacre
of the innocents, I'd pause just to consider the confusion
of your imagery."
—Christopher Fry,
The Lady's Not for Burning, Act III

The linguist Benjamin Lee Whorf is fashionably out of favor at the moment. Questionable and quirky as his methods may have been at times, he evolved radical insights about language that influenced generations and still do. As happens so often with black sheep, people tend to lump all his ideas together and discard them, or reconstitute and rename them. When I first read Whorf in college, I was reading turn-of-the-century physicists and also the Imagist poets. In an art of the exploded moment, the Imagists were enchanted with relativity, what it means to know something, how it looks to feel something. All three arose in the same era, and they cross-pollinated in fascinating ways.

Instead of navigating the shoals of current linguistic theories, which others do so well, I'd rather return briefly to Whorf and la langue out on parole. Half guidebook, half primer, his famous *Language, Thought and Reality* borrowed many ideas from the work of Gauss, Born, Heisenberg, and Planck, and appliquéd them onto his sense of language as trickery and illusion.

Early in the 1800s, Carl Friedrich Gauss had suggested a statistical approach to scientific observation: replace a single reading with a group of readings plotted on a curve. What made Gauss's curve novel is that in the center one does not find a reliable average, only an uncertain center in a range of uncertainty. Neatly put, it's "a concept of continuity, with the ideas as relative locations on a continuous medium." But the quotation is from Whorf, and refers to the spectrum of indelible, though barely discernible, nuances that lead a speaker from one idea to another. Whorf pleaded for a term that would fit this conceptual glide better than "connection" or "association" does. (He finally resorted to "rapport.")

Ideas reverberate, he explains; they do not jump. If we are *up* in a sentence, say, and motion takes place, then "the idea 'up' is a sort of neighborhood, and we are leaving that neighborhood." Whorf was hoping, through the porte cochere of one language, to discover the stock of conceptions common to all languages.

Whorf's ideas seemed daring for his time, but he was very much in keeping with the trend in physics. His creed—that the observable world is as indistinct as the observer is inexact—was a faith shared by the physicist Max Born, who argued that the perceiver is irrevocably knotted up with the perceived, and life's ur-units lie far beyond the radar net of our senses. Meanwhile, Werner Heisenberg was formulating his uncertainty principle: that nothing ever stays exactly the same; what we perceive from moment to moment is a rough estimate of the place and speed; and, therefore, no event can be described with certainty. With so much scope for error built in, any vestigial hope of absolute knowledge went out, even though science does arrive at answers. So the parallel movement in physics was well founded and under way when Whorf produced his theory of linguistic relativity.

Essentially, the Whorf hypothesis has three tines: 1) that the language we speak determines how we react to the world around us, 2) that the Western way of life is a product of how we talk, and 3) that language does what science does: dissect and organize experience. He was particularly good at ransacking satellite disciplines for what might be germane. His knack for finding subtle connections was uncanny; it helped confirm his intuition that language, like science, is a set of iron-fisted treaties. More than once, Einstein wryly observed how convenient it seemed that reality played by rules invented by human scientists.

Whorf also responded to the wave-particle theory of Max Planck. As a corollary to his quantum theory, Planck described light (and other radiation) as a figure-ground, in which the particle was simultaneously part of the wave. Light could be thought of as both the bullet and the sound of the shot. Whorf saw the linguistic and metaphysical implications. His wholehearted acceptance of Planck's, and later Einstein's, theories provided him with exactly the kind of context he needed to unravel Hopi grammar, in which, he claimed, words like *speed* and *rapid* did not exist, their closest parallel being degree of intensity; all events, including thought, had repercussive effects; and a superabundance of verbs conveyed life's interlaced processes. He may not have actually visited the Hopi, relying instead on a written grammar, which conveniently fit his ideas. He tinkered with the notion of chemically compound languages (such as Nootka and Shawnee) versus mechanically mixed languages of Indo-European origin. "We are inclined to think of language simply as a technique of expression," he said, "and not to realize that language first of all is a classification and arrangement of the stream of sensory experience which results in a certain world order . . ." Since all science parasitically employs language, Whorf reasoned, we are bullied by our syntax into accepting as scientific currency facts we might otherwise ignore, a case for defiant inquiry, not hand-me-down resignation. He argued that syntax often requires that we name a physical thing by splitting it into a formless item and a form (introducing a container or shape—say, "glass of milk," "piece of bread," "moment of time"—when the only issue is its contents). And that the sense of time our syntax allows us, devoid of eccentricity, becomes a quantity that can be "saved," leading, among other things, to the "speed syndrome" and our preposterous view of nature as monotonous.

In his vivisections of language, Whorf was seeking an alternative to the self-imposed, arbitrary, mechanistic logic he found ruling science and society. Now and then he claimed to be angling for meaning, that compound ghost of all abstract nouns, but would have settled for the universal bank of conceptions that lured him from the start. Our habit of describing spatially what is not spatial, for example. Consider:

> I "grasp" the "thread" of another's argument, but if its "level" is "over my head" my attention may "wander" and "lose touch" with the "drift" of it, so that when he "comes to his point" we differ "widely," our views being indeed so "far apart" . . .

Following along these lines, I think Whorf would have liked *Metaphors We Live By*, by George Lakoff and Mark Johnson, a linguist and a philosopher who also engage in that search for a bank of conceptions. Lakoff and Johnson argue that, based on physical experiences, the brain forms metaphors to understand "one kind of experience in terms of another," as new metaphors create new realities.

So, for example, since our body is a container, we imagine the world full of containers, and continually try to open things and peer into them. Thus we *open* a conversation, or can be *immersed in* a problem, we *fall in* love, and *come out of* depression, we desire a *full* life, we're *in* the choir. Some other habits of mind include seeing as touching ("I can't *take* my eyes *off* her"); thinking of ideas as a natural resource ("He *ran out of* ideas"); understanding things in human terms ("Our biggest *enemy* right now is inflation"); regarding big as important, more, or better ("Those are *big* shoes to fill"); envisioning love as a journey ("I don't think this relationship is *going anywhere*") or as a physical force ("They *gravitated* to each other"); equating closeness with influence ("Who are the men *closest* to Khomeini?"); perceiving objects as coming from substances ("Our nation was *born out of* a desire for freedom"); regarding emotions as physical events ("His mother's death *hit* him *hard*" or "That *blew me away*"); equating understanding and seeing ("It *looks* different from my *point of view*"); and many more.

Returning to the neighborhood of *up*, we organize our basic beliefs about the world with spatial metaphors. In Western culture, happy is up and sad is down, conscious is up and unconscious is down, more is up and less is down, future events are up, high status is up, good is up, virtue is up, rational is up, active is up, passive is down, gloom is down, and so on. (Lakoff and Johnson don't mention that male is up, female down.) They freely admit that their astute, fascinating, wholly persuasive compendium of some of the brain's favorite metaphors will inflame Chomskyites, and anyone else invested in what the authors decry as "the myth of objectivity"—that is, the belief that things have an innate truth that's independent of us. They're also opposed to the "myth of subjectivity," but not as fervently, favoring a blend of the two. All truth is negotiated truth, and the brain negotiates at the level of metaphor, where our environment and culture impose categories of belief and we define and redefine our self through continually revised personal metaphors that highlight, order, and ultimately make bearable some aspects of our lives, while hiding other aspects on purpose.

For centuries, Western scientific, philosophical, and linguistic tradition has taught us that the world is littered with absolute truths waiting to be uncovered. Science does work, this tradition reasons. We humans are clever at self-soothing. But metaphor isn't just decorative language. If it were, it wouldn't scare people so much. I was reminded of this recently, while serving as a judge for a literary prize. Our genre was nonfiction and included books of history, politics, philosophy, science, belles lettres, psychology, and memoir. My two co-panelists found the prose in all the books I recommended too "complex," "self-conscious," and "mannered." They dismissed them for "calling attention to themselves," and felt viscerally that *all* prose should be "clear," "transparent," "invisible."

Colorful language threatens some people, who associate it, I think, with a kind of eroticism (playing with language in public = playing with yourself), and with extra expense (having to sense or feel more). I don't share that opinion. Why reduce life to a monotone? Is that truer to the experience of being alive? I don't think so. It robs us of life's many textures. Language provides an abundance of words to keep us company on our travels. But we're losing words at a reckless pace, the national vocabulary is shrinking. Most

Americans use only several hundred words or so. Frugality has its place, but not in the larder of a language. We rely on words to help us detail how we feel, what we once felt, what we can feel. When the blood drains out of language, one's experience of life weakens and grows pale. It's not simply a dumbing down, but a numbing. There's a big difference in how we experience "This rose smells good" and "This rose smells like hot cocoa" or "This rose smells like violin resin."

"Invisible prose only!" rules out the sparkling style of such writers as Virginia Woolf, Thomas Carlyle, Walter Pater, John Ruskin, Thomas De Quincey, Djuna Barnes, Vladimir Nabokov, Edith Sitwell, Herman Melville, James Joyce, Rainer Maria Rilke, Jorge Luis Borges, John Donne, Oscar Wilde, Marcel Proust, E. B. White, Sir Thomas Browne, William Faulkner, and Samuel Beckett, to name only a few. For those writers, vivid prose, and the visionary mind it evinces, rich with speculation, insight, and subjectivity, *is* the craft and offers a unique caliber of truth. Is there any other art form one would praise by saying it's "invisible"? By definition, art transcends the ordinary, calls attention to itself, and offers virtuosity as its calling card. One that makes it possible to do what metaphor does so well: illuminate some of what can't be wholly understood.

Artists don't create lavish metaphors out of nothing; they perceive the world lavishly. But we all use countless metaphors every day, unconsciously, to provide an important hinge between feelings and ideas. Just as raindrops need dust motes to condense around, metaphors need perceptions. Because the mind inhabits the visceral body, and relies on the abstraction of language, it needs a way of embodying thought, of making ideas sensible, of probing the world even when the body is resting. Through metaphor, thought becomes action that can be staged in the mind's eye. We do it so fast we rarely notice that it's an elaborate what-if game. What if I picture a startled crow that suddenly flies to a tree limb, what if I track it in separate eye-gulps as it angles up steeply, one rigid flap after another, seemingly frozen in space. What if I note its alarm, speed, and angle at the same time. What if, at a later date, I arrange planks of wood in a way that resembles the crow's dash from ground to sky. I might call it a *flight* of stairs. Just as I might see and use a ladder, pausing at each rung to haul myself a step higher, and later describe a woman *laddering* her stockings when she puts them on.

This requires noticing many details, choosing the definitive ones, and selecting only those shared by something else. A mixed bag of attending, sensing, and filtering. We do it all the time, effortlessly. Some do it more elaborately. Some paint it with emotion. Then the what-if game becomes really complex. What if I remember how I felt when my father yelled? What if I insinuate myself into that memory and dredge up the sensory feelings? What if I sift through all my touch experiences for a similar sensation? What if I stop at the time I snagged my finger on a sharp grating board while making coleslaw? I might well wince and say his voice *grated*.

Not everyone experiences words as living scenes. Most people slur over the details. I wince whenever I see or use the word *grate*. As I thought and typed the word *slur*, I saw a floating nylon mesh that blurred as it moved. And typing *blurred* I saw frosted glass. And typing *frosted* I saw a storybook face of Winter bearded in ice and snow. I don't sense such things urgently, or even in the forefront of my thoughts, but always. Like a mental whisper or fleeting glimpse. Words are technicolor slides the mind quickly views before using them as symbols. Some people are naturally more aware of the slide show behind the curtain of language, but everyone can learn to pay attention to it. Why the curtain? For the same reason that much of the brain stays unconscious. Because otherwise we'd meddle and screw things up.

Instead we encourage the self-deception at every turn. We can and do see the spaces between a bird's flaps. Consciousness is staccato, not fluent. We perceive in tiny packets of information. Our attention is easily perforated. But we need the world to seem fluent and intact, otherwise it would be unbearable, not to mention unpredictable. So we caulk up the gaps and insert shims. We talk easily, amiably, without thinking about the nuggets of social history or sensory truth words contain.

Metaphor is one of the brain's favorite ways of understanding the this and that of our surroundings, and reminds us that we discover the world by engaging it and seeing what happens next. The art of the brain is to find what seemingly unrelated things may have in common, and be able to apply that insight to something else it urgently needs to unpuzzle. It thrives on analogy. To some, being aware of that process is exhilarating, to others it's scary, depending on one's need to believe in absolute truth, and deny the extent to which the brain uses metaphor, often imperceptibly, relying on what we do know to illuminate what we don't.

CHAPTER 29

The Color of Saying

Poetry is a dream dreamed in the presence of reason.
—Tomasso Ceva

All language is poetry. Each word is a small story, a thicket of meaning. We ignore the picturesque origins of words when we utter them; conversation would grind to a halt if we visualized crows whenever someone referred to a *flight* of stairs. But words are powerful mental tools. We clarify life's confusing blur with words. We cage flooding emotions with words. We coax elusive memories with words. We educate with words. We don't really know what we think, how we feel, what we want, or even who we are until we struggle "to find the right words." What do those words consist of? Submerged metaphors, images, actions, personalities, jokes. Seeing themselves reflected in one another's eyes, the Romans coined the word *pupil*, which meant "little doll." Orchids take their name from the Greek word for testicle. *Pansy* derives from the French word *pensée*, or "thought," because the flower seemed to have such a pensive face. *Bless* originally meant to redden with blood, as in sacrifice. Hence "God bless you" literally means "God bathe you in blood." The snub of a *cold shoulder* originated in Europe, during the Middle Ages, when people who overstayed their welcome were served cold beef shoulder (rather than hot food); after a few cold meals, guests got the message. We say "windows" because Norsemen kept their doors closed in winter, relying on a ventilation hole (or "eye") in the roof. The wind played through it expressively, and it became known as *vindr auga*, the "wind's eye," which the English changed to "window."

We inhabit a deeply imagined world that exists alongside the real physical world. Even the crudest utterance, or the simplest, contains the fundamental poetry by which we live. This mind fabric, woven of images and illusions, shields us. In a sense, or rather in all senses, it's a shock absorber. As harsh as life seems to us now, it would feel even worse—hopelessly, irredeemably harsh—if we didn't veil it, order it, relate familiar things, create mental cushions. One of the most surprising facts about human beings is that we seem to require a poetic version of life. It's not just that some of us enjoy reading or writing poetry, or that many people wax poetic in emotional situations, but that all human beings of all ages in all cultures all over the world automatically tell their story in a poetic way, using the elemental poetry concealed in everyday language to solve problems, communicate desires and needs, even talk to themselves.

When people invent new words, they do so poetically—arguments have *spin*, a naive person is *clueless*. In time, people forget the etymology or choose to disregard it. A plumber says he'll use a gasket on a leaky pipe, without knowing that the word comes from *garçonette*, the Old French word for a little girl with her hymen intact. We dine at chic restaurants from porcelain dinner plates, without realizing that when smooth, glistening *porcelain* was invented in France long ago, someone with a sense of humor thought it looked as smooth as the vulva of a pig, which is indeed what *porcelain* means. When we stand by our scruples we don't think of our feet, but the word comes from the Latin *scrupulus*, a tiny stone that was the smallest unit of weight. Thus a scrupulous person is so sensitive he's irritated by the smallest stone in his shoe. For the most part, we are all unwitting poets.

It is just one habit of the brain: finding relations between things, especially between seemingly unrelated things. Seemingly, because all things are related in the web of life on Earth. True, quartz is different from a member of a college swim team, but they share many features. Not to mention that the word *quartz* began with someone thinking of it as a "siren" (the etymology of *quartz*), an enchantress who lured men with a song of colors, liquid as light, but deadly as rock. If pressed, one could find ways to relate quartz and a member of a college swim team. Perhaps through water—the pool being contained water with a few chemicals tossed in, as is the man for the most part, as is the quartz, through which a fluid (light) still pours. Or the changing face of the man, the changing faces of the quartz. Or that each began as a miniature version of itself, in a dark recess, and grew large. We rarely think of crystals growing, but they do, and they grow in a way we associate with babies growing into lawyers or heliarc welders—otherwise we wouldn't use the term *grow* for both.

Sometimes I think we mainly invent words to help picture ourselves: in metaphorical mirrors. We humans are easy to know, but hard to know well. Because we plant and cultivate, we can imagine a seed planted in the womb and growing, see children as a sort of crop. Because we build machines, it's easy to depict the body as mechanical, or even as a factory. Because we use computers, we envision the brain as "hardwired," our socializing as "networking," our skills as "software."

Despite our best efforts, the closer we look at anything, be it wildflower or fever, the scrappier language becomes. It fails where we need it most, at the outskirts of mind, memory, and emotion. Poets solve this problem by fusion (metaphors), bridging (similes), and other devices. But whole cultures do it, too. "Surfing the Internet," for example. Or computers having "viruses." In time, those words will pass through the lips of countless humans, and evolve into other words just different enough to disguise the original poetry behind them. The art of the brain is to use poetry to navigate the world. We breed symbols, we speak fossil poetry. Even so, how would you describe the sun to a blind man? One wordsmith knew.

CHAPTER 30

Shakespeare on the Brain

[He understood] . . . the quality of the real universe, the divine, magical, terrifying and ecstatic reality in which we all live.

—C. S. Lewis

Planning a small theme garden with roses named after characters in Shakespeare's plays, I was surprised to find dozens of choices, from blood-red Dark Lady to pale, blush-colored Juliet. I doubt Shakespeare would mind his characters being transmogrified into pricey plants. But the Shakespeare Fishing Tackle company? Shakespeare Needlework Kits? Shakespeare Magnet Poetry? Shakespeare, New Mexico, "the West's most authentic ghost town"? Then there are all the restaurants, motor lodges, bookstores, hair salons, household products, and paraphernalia named after him and his works. The admiration of countless readers he'd welcome. After all, a favorite argument in his sonnets is that his beloved will live eternally in the poems, which will be eternally loved. But Shakespeare foods and toiletries? Online Shakespeare sites, where visitors vote for their favorite play? Entire libraries devoted to the mysteries of his life and the majesty of his works? Those would surely give him pause.

What I think would delight, frighten, and fascinate him at the same time is the simple, universally accepted truth that when it comes to artistic genius, he stands alone. There's Shakespeare, followed by a very large gap, and then all the other English writers who have ever lived. No one could write like him, not even such renowned stylists as Browning and Nabokov. Something about his brain was gloriously different.

Familiar enough to illuminate the human condition in recognizable, entertaining, and profound ways, but different enough to do it in ways and words no one else could achieve. Something about the radar net of his senses was different. Something about his ability to combine seemingly unrelated things in a metaphor's alchemy was different. His ability to juggle many swords of insight at the same time was different. In truth, the people of his era had a very small vocabulary; ours is exponentially larger. But his gift didn't require more words, because words, being human made, can't begin to capture the experience of being alive or the complex predicaments even simple people get into. Words are small shapes in the chaos of the world. They're unwieldy, sloppy, even at their most precise. Nothing is simply blue. No one just walks. Words fail us when we need them most. They fall into the crevasses between feelings. If we make them overlap, then we can cover some of those spaces, and that's traditionally what writers, especially poets, do. A metaphor is hypergolic, like nitroglycerin. It takes two otherwise harmless things, smacks them together, and creates something explosive. Instead of needing a vocabulary word for every single thing and experience, we use the words we have in new ways. How clever of the brain to find such an enchanting solution.

The thing is, Shakespeare did this more accurately than anyone. He pictured rumor as a "pipe/Blown by surmises, jealousies, conjectures." He described a kiss as "comfortless/As frozen water to a starved snake." In *Richard II*, his king metamorphosed into "time's numbering clock;/My thoughts are minutes, and with sighs they jar."

It's not just that his senses were unusually keen, though they were, or just that he was a patient and detailed observer of human nature, though he was. He must also have possessed a remarkable general memory, the ability to obsess usefully for long periods of time (*usefully* being the operative word here; he probably could obsess uselessly, too), a superb gift for focusing his mind in the midst of commotion, quick access to word and sense memories to use in imagery, a brain open to novelty and new ideas. His personality must have included these traits, too, because they're essential to all creativity: perseverance; resourcefulness; willingness to take risks; the urgent need to make exterior his inner universe; the ability to live not only his own life but also the life of his time; a mind of large general knowledge and strength that could be drawn to some particular phenomenon or problem; the capacity to be surprised; passion; the innocent wonder of a child made available to the learned, masterful adult. That all these qualities, and many more, might combine to produce what we refer to as a moment of *inspiration* is one of the great mysteries and triumphs of mind.

His was a dangerous era in which to be a playwright at court. Elizabeth I and James I attended private performances of his plays, and he had to be tactful about what he said. I'm sure he would have understood the embroidered pillow I once saw in a shopwindow in Palm Beach that read "Be careful. The toes you step on today may be connected to the butt you have to kiss tomorrow." But spontaneity bounded by restraint is the stock-in-trade of creativity. It's one of the brain's favorite ways of creating art. Thus paintings have frames, symphonies and verse forms have strict rules. The restraints aren't always political. Sometimes the medium has built-in limits—oil paint mixes and dries in characteristic ways. As technology changes, limits sometimes change, as when oil paint became portable. Some restraints are ordained by society.

For example, in Shakespeare's day, arranged marriages were still the norm, but many people started objecting to them. His plays are filled with commotion over whom to marry, the right to choose a mate, and complaints by characters who'd prefer a love match. The best known of them, *Romeo and Juliet,* was a classic told in many cultures and genres when Shakespeare decided to tell the story yet again, doing what Leonard Bernstein and his collaborators did with *West Side Story*—adapting a well-known, shopworn tale to contemporary dress, locale, and issues.

They created a younger Juliet than in other versions, allowing a beautiful, thirteen-year-old Veronese girl to encounter the embodiment of her robust sensuality: a boy who is passion incarnate, someone in love with love. "Love is a smoke made by the fume of sighs" [I, i, 190], he tells his friend Benvolio. Romeo is a bolt of lightning looking for a place to strike. As Romeo declares: "Juliet is the sun" [II, ii, 3]. When he meets Juliet the play's thunderstorm of emotions begins. All versions of the story hinge on the rivalry between two noble houses, and the forbidden love of their children. In Shakespeare's, chance, destiny, and good playwriting ordain that they shall meet and become "star-crossed lovers" with a sad, luminous fate. The use of lightning and gunpowder images throughout the play keeps reminding us how combustible the situation is, how incandescent their love, and how life itself burns like a brief spark. Full of tenderness and yearning, their moonlight balcony scene contains some of the most beautiful phrasing ever written, as they sigh for love under the moon and stars, most alive in a world of glitter and shadow.

A director friend told me recently that while watching news coverage of war-ravaged Bosnia one evening, he began thinking differently about Romeo and Juliet. Suddenly he realized that the play isn't really about star-crossed lovers. It's about what happens to children in a culture of violence; and that's how he directed it at the Shakespeare Festival in Kansas City. Because Shakespeare excelled at constructing a house of cards, whose walls are intersecting planes of meaning that support one another, both interpretations ring true.

Another angle on Shakespeare's brain is that he wasn't good at inventing plots. He elaborated them cleverly once he had them, but for the most part he borrowed plots from historical sources. As I understand, sadly, plotting requires a special cast of mind. Give me a ready-made plot and I'll have fun elaborating it. Ask me to make phrases until the cows come home, and I'm happy. Invite me to describe a gesture or set a scene or develop an idea or explore someone's psychology, and I'll roll up my sleeves. But ask me where the person crossing the room just came from, and I have no idea. It's a mechanically different process. To do it easily takes a brain tilted in a slightly different way. An example of this is the potboiler, a one-dimensional book that has nothing to do with literature, although its plotsmanship may be brilliant.

Style flows from the sediment of one's personality. It's hard to know what Shakespeare really believed because his different characters contradict one another. I've written dramatic monologues, and a verse play, and there are reasons poets find them compelling to write. Here are a scant few: Sometimes one writes passionate soliloquies about feelings one is only testing out, to probe them, to see what they would feel like, painfully, enthusiastically—but not irrevocably. For reasons I won't even try to explain, writers often feel the need to be private in public, and taking on the mask of a dramatic monologue does that well, as a form of ventriloquism. One can splurge on emotions like anger or unrequited love, reveal real and imaginary crimes. One can say politically dangerous things. It gives one license to be licentious. Yet one can disown the feelings as mere inventions, the ravings of a buffoon created for stage effect.

Style is also a penetration of the world, an adventuresome safari through daily life. The drafter of "O that this too too sullied flesh would melt" has probably witnessed the slow putrefaction of pork as well as ice melting on a pond in Warwickshire. The line is a blurt, an educated man's yearning to be nobody (rather than a young prince obsessed with his royal father, whose image he carries about with him). The energy of that statement comes from a Shakespeare who has opened his own soul to the possibility of evaporation.

A self is a frightening thing to waste, it's the lens through which one's whole life is viewed, and few people are willing to part with it, in death, or even imaginatively, in art. Shakespeare was a master of empathy and surrender, who gave himself to the human dramas he saw daily, figuring out how he would feel if doing or saying certain things. He became a crowd. I don't imagine this is something one turns on and off very easily, or suddenly decides to try; he probably imagined himself as others often, secretly, from childhood on. Why he would have found escape from self necessary, or fun, is not for me to speculate. Most people won't be pried loose from the single word *I*, which is, in the end, our only possession. He savored human phenomena, humans as intriguing sensory suits that he borrowed now and then. With equal measures of grandiosity and humility, Shakespeare would slip out of his self and into another's, then quiz his senses, and cast the feelings and sensory information into his best language. I wonder how many Autolycuses he knew personally before he came up with the definitive phrase "a snapper-up of unconsidered trifles."

Shakespeare had the courage of his diffidence. He soaked things up and brooded on them, before allowing them to fill the mouths of Macbeth, Portia, and Prospero. Of course, to be so giving in your response to life, you have to be mighty sure of yourself, aware that you will always be able to whip out a unique phrase that will displace all the efforts of all the other authors who don't achieve anything like sixty-odd pages in *Bartlett's Familiar Quotations*. You also have to be willing to experience extremes of emotion, no matter where they lead, even as far and low, say, as that moment in *King Lear* when, grieving over the body of Cordelia, he laments: "Never, never, never, never, never." Five *nevers* is a lot. The words function as individual guillotines, and there's no question about the anguish felt in the writing of them. It is the numb bleat of an author who diagnoses individual doom.

At the end, Hamlet beseeches Horatio to report him aright in the world, to paint an accurate picture; he knows no celestial scribe is going to do the job for him, or the nonexistent media of the day. Word of mouth is all, and, as Shakespeare foretold in his sonnets, it has served him well over the centuries. It's not enough to say that a certain play of Shakespeare's is full of quotations, or even, as George Bernard Shaw said of *Henry V,* that it's a national anthem in five acts. The mind at work in all the plays has a unique idiom. Something was wonderfully different about his brain.

The texture of his imagination favored certain rhythms and patterns. "The bank and shoal of time" is one of his treasured rhythms: the one and two of three. When you think about it, that's a convoluted way to build a metaphor. In "the slings and arrows of outrageous fortune," for instance, items owned by an unstable human archer (slings, arrows) imply the calamitous actions they're capable of. And then the phrase says: "Okay, now swap fortune for archer." A listener hears the first part, allows the items to conjure up the man, pictures the archer, then is surprised by the appearance of something abstract—fortune—which has now acquired familiar and scary human characteristics. All that in a phrase heard fleetingly, and yet our brains follow the trajectory of the imagery without a sweat, and we're powerfully moved by it. Shakespeare has managed to endow a superstitious force with a ferocity we have seen, or perhaps even possessed.

There are some rhythms used successfully by other poets— Browning, say, or Swinburne—that Shakespeare didn't seem to care for at all. Iambic may be the natural rhythm of someone walking, but how many people can explore a mood in blank verse? His thoughts had a unique cadence. Some of his characters talk to themselves in sonnets. It's appealing to use the straitjacket of a verse form to organize intense emotions, and I'm sure he worked hard at word choice, as all professional writers do, especially ones with appalling deadlines. But, nonetheless, his mind seemed to entrance itself with verse. I wonder if his favorite rhythms didn't work as a reliable sort of incantation, which focused the mind, disrupted the mundane hodgepodge of thoughts one needs to buy bread or shoot the bull with pals, and spirited him away to another mental kingdom.

Did Shakespeare know how different he was? Probably so. Even in idle chitchat, people report on what they feel and see, revealing how the world touches them and the realm of their sensibility. He would have known how alien he was. How human in a hundred familiar ways, but also how different. It would have been both his privilege and burden to be extraordinary. More of everything. More hearing, seeing, feeling, smelling, more imagining, and per- haps more hurting, thrilling, angering. What was the texture of his imagination? Would mapping his brain have told us? If he were alive today, could scientists begin to explore the mountains of his mind? What would they find? A rare bounty of connections among neurons, which makes it easier to link unrelated images from dif- ferent parts of the brain: the stuff of metaphors? Since our brains sprout too many connections, and then heavily prune, maybe his brain did a lot less pruning, maybe his neurons stayed bushy.

Recently, longing yet again to know who Shakespeare the man really was, I phoned the Shakespeare specialist at an Ivy League university and asked her advice on which biography to trust.

"Oh, I'm not interested in the man or his work," she said.

"I see," I said, pausing until the shock wore off enough that I could continue speaking. "Well, if you're not concerned with the man or his work, then what aspect of Shakespeare *are* you interested in?"

"The political response to the canon," she said decisively.

"And that's what you teach?"

"Of course."

"Not the man or his work?" I asked this in as neutral a tone as I could, since I sincerely wanted to know her answer.

"No!" she said impatiently, and her voice had a condescending edge.

Clearly, if I wanted any information from her, I would need to apologize for my aberration. "Well, perhaps you'd humor me, you know how writers are—I'm enchanted by the work and would like to learn more about the man." She suggested a biography that relied heavily on legal documents from Shakespeare's time. It was a useful suggestion. But when we hung up, I felt enormous pity for the students who would find Shakespeare's work reduced to the tidy annals of political theory, and a profound visceral sense of desecration. Not because Shakespeare was a god. If anything, he risked being more human than most. Because he was a natural wonder.

THE
WILDERNESS
WITHIN

(The World We Share)

CHAPTER 31

Oasis

. . . the spirit of inhabitable awe . . .
—Edward Hirsch,
The Demon and the Angel

The armor-plated lump of impulse and self-counsel we call the brain began in ancient seas as a bulge of nerves atop a spine. But long before that, blue-green algae ruled the planet. Their gift was the cell, a microscopic circus of fuel and volition so successful it's still the basis of every life-form from cougar to walnut. Their genius was inventing photosynthesis. Around 2.4 billion years ago, they began building solar power plants behind their walls and digesting their surroundings, excreting oxygen, a poisonous gas, in the process. This waste bubbled up in translucent ribbons. Over time, blue-green algae sheathed the planet and oxygen fizzed through the oceans, saturating them. Then, having nowhere else to go, the bubbles rose free, breathing life into a slaggy sky, whose sour cloudbanks thinned as the blues appeared. Hydrogen ballooned away into space, while heavier oxygen stayed home.

Meanwhile, over millions of years, evolution tinkered with creatures immune to oxygen, including some willing to cooperate by pooling their DNA. None of these life-forms had brains. And yet, in retrospect, they seem clever, inventive, ambitious, resourceful. Surely a word as long and low as *accomplishment* requires a brain of some sort? Otherwise, dare we admit that our exalted intelligence may be humble, a know-nothing kit chanced upon by random mutation, that we're bastards of witless one-celled organisms? In every neuron and flake of skin, we still resemble those pioneers. I find the career of such unilluminated creatures beguiling, even awesome, but I may be unusual in that. No matter how politely one says it, we owe our existence to the farts of blue-green algae.

If life doesn't require a brain, why did our brain come to be? Many sea creatures had to wait for food to brush up against them or enter their flailing orbit. Daring others, guided by smell, began to roam, hunting food, pursuing mates, scenting danger. Smell compounded the possible, fed hankerings, heightened peril, and daily life became more complex. Because it made short work of detecting predator and prey, smell thrived, was joined by other senses, and in time that tissue swelled into the cerebral hemispheres. Smell became so successful, there was little need to change it, and it still works today pretty much as it did in the ancient seas. We move in oceans of air, but odors still must melt into a solution before our nose can absorb them and tell the brain we're smelling smoke or honeysuckle. An assembly of companion senses linked to a motor center went about its chores, and still does. Nodes, bulges, and clumps appeared, as the brain grew bigger and more elaborate—the thalamus to integrate the senses; the emotional amygdala; the hippocampus aiding in memory; and the hypothalamus, regulating the body's basic needs and appetites. Grouped as the limbic system (because they happen to lie adjacent to one another), this second unconscious dynamo aroused ghostly feelings we rarely noticed.

Fast-forward through eras scored with key themes, as an opera of life-forms auditioned for future roles, plumed and befrilled, modeling new body parts and costumes. Some tried a big brain, others a little one; and both triumphed. Small-brained animals thrived in habitats that didn't require much ingenuity. Big-brained animals braved fickle habitats by improvising and puzzling things out. They prevailed because the opposite sex prefers survivors, whom they mate with, passing on winning genes. Animals don't evolve just to adapt to the environment. To scatter their genes, they must outbreed their neighbors. As Darwin understood, survival of the fittest really means survival of the sexiest.

Consider scrub grasshoppers, rare insects found on Highland

Hammock in central Florida. Among the males, genitals evolve fast to tantalize desirous females. Some males develop a few little extra prongs on their genitals, the females prefer the prongs and mate with those grasshoppers, and their offspring carry genes for prongs. What's so sexy about prongs? Well, for one thing, they're hard to ignore. If a female is absolutely certain she mated, she can stop wasting time and energy in courtship. So if a male excites her with lots of sales talk or a fresh design, he'll succeed in mating with her, and she'll reject other suitors. Needless to say, this has led to all sorts of genital trinkets, including bundles of hair, little hooks, bristly reversible sacs, and knobs that point in opposite directions. They're adaptive only in the sense that by satisfying the female's yen they ensure mating, but they're not useful, just decorative. Some of our own evolution may be similar, from lip shape to mind games. We evolved the sexiest brain we could, and, when the poker chips are down, *sexy* means built for survival. Lifesaving new behaviors appear by chance, percolate through a population, and are passed along to offspring.

Stressed by landscapes littered with hazards and the sublime obliquities of desire, our brain grew a new frontier all its own, the neocortex, a wrinkled metropolis where billions of brain cells configure thought. Many sprout wiry connections that crisscross one another and sometimes run far afield to contact other neurons, and so weave the bolting fabric of the mind. Deep below, the old brain keeps toiling like the fabled ogre in the dungeon, doing its chores, making mischief, vexing reason and courtesy whenever it can.

Because of the neocortex, we became highbrows, as an eruption of new brain tissue pushed out the head bones. Endlessly mutating, the brain tried out many strategies, including self-awareness and abstract thought, while holding on to everything that worked well for learning and memory. Combining skills old, new, and borrowed, our brain eventually swelled to what it is today, roughly the size and shape of "an old crumpled boxing glove," as neurologist Richard Restak picturesquely puts it.

Life vexed and taught us. We hunted a wide variety of food, but much of it fought back, and some evaded us, grew out of sight or reach. Meanwhile, better-equipped predators were busily stalking us. Rarely could we outmuscle our prey, so we tried to outwit it. We devised traps and crafted weapons. Our diet was complex, our habitat varied, so we needed to be nimble, inventive, keen-eyed, easily startled, flexible, naturally curious. Because most of our food could outswim or outrun us, we learned the slow art of anticipation, and the fast art of sizing up a problem and deciding what to do. We needed cunning and intuition to spar with rivals and juggle social hierarchies. Our new brain learned to cooperate and to judge, to laugh and to worship, to love and to ridicule. It became an ace illusionist, and the ultimate voyeur, watching itself watching itself at work while also watching the world.

It's tempting to think of evolution as a hereditary life jacket, something we receive, not rework. Because each generation stirs the pot, changing its physical, social, and psychological environments, we've been tinkering with human evolution all the time. Sometimes the tinkering is planned—cities, industry, pharmaceuticals—sometimes it's a dangerous by-product of well-meant progress (pollution, X rays). It's shocking how much garbage a parade of good intentions can leave behind. There's a certain thrill about that for some people—we can radically alter the world to ease suffering, but also just to amuse ourselves, or to show off. Yet all the while we continue to evolve and adapt. Our brain is simply one experiment life is running. There are many others, equally unlikely. We don't know the outcome of these trials, we don't even know all the ingredients. Meanwhile, we try to keep an eye on the whole shebang, from prairie vole and ozone layer to skunk cabbage and humans, astonished by the long shot of our being here at all.

CHAPTER 32

Conscience and Consciousness

We *are* something we can't *know*.
—Michael Eigen,
The Psychoanalytic Mystic

A re we the only conscious creatures on the planet? To begin to answer that, we need to define human consciousness, and as I was saying earlier, conflicting theories abound. The philosopher David Chalmers, of the University of Arizona, draws a distinction between the "easy" problems of consciousness, which means understanding cognitive functions like storing memory and paying attention, and the "hard" problem—which is why any of that should create consciousness. But even the easy problems are hard, and the quest to solve both is focusing some of the best brains in science, psychology, and philosophy.

Steven Mithen analyzes the rise of the brain, from a waxy gray ball possessed by our distant mouselike ancestors 65 million years ago, to an evolving rotunda atop many mammals, some primates, and several hominids, to the mind's pulsarium we now know and love. For most of that drama, he ventures, the brain grew highly specialized intelligences cut off from one another with an absoluteness hard for us to fathom. Spared mutual awareness, those intelligences were unable to confer and intersow skills or observations. A natural sense helped them forage, read the weather, track prey. A technical aptitude led to working the world's materials and creating such useful objects as stone axes. A powerful social intelligence guided them through the shifting narrows of relationships. Each

lens widened their outlook and chance of survival, their viability as
a life-form, and required hard brainwork to master. Unfortunately,
with a cognitive barrier keeping these special intelligences apart,
craft learned by one domain couldn't solve another's problems. That
loose assembly survived for millions of years, but when chance
mutation produced a few outlandish individuals blessed with what
must have seemed otherworldly to their tribe—cooperating intelligences—they thrived, attracted more mates, were better at protecting their young, who inherited their integrative genes, and so
on. Once the different domains knew about each other, they could
cooperate, though also meddle. In Mithen's view, the role of consciousness is to integrate "knowledge that had previously been
'trapped' in separate specialized intelligences." With that reci-
procity of know-how, observation, and experience, a tactic or skill
gained in one mental domain could be recruited by another. What
follows from that bringing together of unrelated things is analogy,
metaphor, and language.

This psychic interfusion would have welled up slowly at first, but
in time acquired an unstoppable velocity, allowing the brain new
flexibility, creativity, and muscle. Just as in the Japanese folk art of
sword making, where metal is heated, folded, hammered, heated,
folded, and heated again, over and over, until it becomes thinner but
stronger, the brain became more versatile as it folded its parts
together and built interplay into its design. Tool savvy could be
shared with a relative, stone working could be applied to other
materials, such as bone and antler. When natural observations
became available to the technical intelligence, one might notice
rainwater pooling in animal tracks and think: *I could use one of those
to carry water.* I don't believe our early ancestors reasoned as we
might: *Sun-baked earth can hold small amounts of water. I want to
carry water. I'll take earth, sun-bake it, and fill it with water. Then I can
carry it with me.* But even the simplest mixing of intelligences
requires thinking about what one is doing, or so-and-so's behavior,
or a new tool design—or all of them together. Mithen argues that
with this "cognitive fluidity," art, religion, complex thoughts, and
thinking about thinking became not only possible but inevitable.

The cognitive scientists Gilles Fauconnier and Mark Turner believe in Mithen's cognitive fluidity (which they rename cognitive blending), though they differ with him on a few points. "The mind is not a Cyclops; it has more than one I," they write in *The Way We Think,* "it has three—identity, integration, and imagination—and they all work inextricably together." They believe language and our modern mind evolved gradually as a changing brain started to blend complex mental spaces. Then language became "a system of prompts for integration."

Here's what I think they mean. In the movie *Sabrina,* it's said of a ruthless businessman: "He thought morals were pictures on a wall, and scruples a Russian currency." To get that double pun, we have to know what murals are, what morals are, what scruples are, and what roubles are, and blend elements from all four to say something else: He was heartless. Yet in an instant, we find it a funny, barbed play on words. That seems to me a remarkable feat. We should be awestruck by the combinative genius of even the most humdrum mind.

Simpler cognitive blends are more analogical, and they make good aphorisms, such as Emerson's "Art is a jealous mistress," "Coal is a portable climate," "Hitch your wagon to a star," "Wit makes its own welcome." If I say "Love's mansion has many rooms," most people can picture it, and understand that rooms vary in size, decor, temperature, and purpose. To make complex blends, you have to be able to shuffle contradictory mental spaces without really being disquieted by the process or even aware of it. That all happens behind the scenes. The mind has busy stagehands who don't need to be seen. Their not being seen must have favored survival, or the unconscious wouldn't play such a large role in our lives. No need to clutter up the stage. However, "that puts cognitive science in the difficult position," Fauconnier and Johnson observe, "of trying to use mental abilities to reveal what those very abilities are built to hide."

According to the philosopher Patricia Churchland, nerve pathways allow us to monitor our body and thoughts, with many layers of self-monitoring. What we experience isn't supernatural, it's the feel of brain tissue at work. She thinks in terms of a compound "set of capacities" able to picture both the physical body and such mental concerns as "our autobiography, what we currently feel about our body . . . where we are in space and time, where we rate in the social order, and the status of our relations to other humans and nonhumans." Faced with hunger, say, the brain imagines a host of physical and social scenarios and quickly chooses a course of action. Then it informs what Churchland calls the Emulator, which considers the consequences and responds with its findings. That may result in a modified plan, whose consequences the Emulator again judges. In time, as this to-and-fro repeats and flourishes, a workable solution becomes clear, the body is informed, and it executes a plan. She sums it up like this: "In brief, the wiring yields self-simulation with respect to the things in the world. But some of those things—at least for those of us who are social creatures—will entail the simulation of other *selves*. What will that organism do if I display anger? What will it do if I chase it? If I try to eat it?"

Gazzaniga's theory of the Interpreter, knitting together miscellaneous brain circuits and systems, is also compelling. The left hemisphere, not content to quietly perceive, insists on storytelling, and that allows us the illusion of being rational and acting with free will. Gazzaniga concludes that the left hemisphere's Interpreter "triggers our capacity for self-reflection and all that goes with it . . . a running narrative of our actions, emotions, thoughts, and dreams. . . . To our bag of individual instincts, it brings theories about our life." The brain engenders a sense of self because "these narratives of our past behavior seep into our awareness; they give us an autobiography."

The philosopher John Searle, of the University of California at Berkeley, sees consciousness as both physical and nonphysical, a state produced by the brain but more than the sum of its many processes. Neuroscientists refer to this as *emergence*. Personally, I prefer a word like *synergy*, since *emergence* is already the technical term for bats swirling into the sky at sundown. An emergence of bats. An emergence of mind. I guess that works as metaphor. "Processes in the brain cause our conscious experience," Searle says. But "consciousness is not like some fluid squirted out by the brain. A conscious state is rather a state that the brain is in. Just as water can be in a liquid or solid state without liquidity or solidity being separate substances, so consciousness is a state that the brain is in without consciousness being a separate substance." He feels consciousness is irreducible not because it's mysterious but because it's so subjective. *My* personal experience doesn't translate into *his* or *hers*.

Stephen Kosslyn suggests that "consciousness is like light that is produced by a hot filament in a vacuum: The physical events that produce the light cannot be equated with the light itself." Consciousness doesn't accompany all of the brain's work; we're rarely conscious of how we make choices and decisions, for example, important as those are to our survival. We're rarely conscious of the patterns we've stored in memory, which allow us to quickly identify things. Kosslyn concludes that consciousness evolved as a supervisor, a quality checker, to discover whether the brain is operating correctly. It can only do that by constant surveillance, an exquisite sensitivity to states of balance, equilibrium, and consistency. If neural activity falters, consciousness detects the change.

Neurons located in different parts of the brain, when responding to the same stimulus, oscillate in synchrony (at around 40 Hz), bound by remembered associations. For Kosslyn, consciousness is related to such a synchrony as a musical chord is to the individual notes that compose it. The chord wouldn't exist without its elemental notes, but it can't be reduced to those notes. The chord may be harmonious or dissonant, the instrument can sound out of tune. Consciousness may not have evolved for that purpose, but slowly, opportunistically, from brain structures designed for other business, which were recruited to construct consciousness. (An analogy he offers is that noses evolved for smell, but they can also be used to hold up glasses.) Researchers measuring electrical patterns have found widespread synchrony when people are aware of their perceptions, but not when they receive information unconsciously. As long as the brain is functioning smoothly, we're absorbed in living, and the brain music plays below our awareness.

But the moment a neural chord starts oscillating wrong, there's dissonance, the arrhythmia bothers us, and we become aware of our conscious state. A sudden strong emotion might produce a jarring note in the chord. In theory, this would lead us to do or feel whatever we must to stabilize the brain again.

These are just a few of many contending theories. Thought-provoking books about consciousness abound, and in the chapter notes and bibliography I list some that I find especially stimulating. Mind you, no one has even proven empirically that we have consciousness. It's just something we intuit from personal experience. Since there aren't any readable dolphin memoirs or translators fluent in chimpanzee, our circumstantial evidence is limited when it comes to other animals. But not nonexistent. All mammals have a cerebral cortex, the probable haunt of consciousness. Many have a basic self, a sense of being here now, separate from the rest of the world; and some seem quite aware of their body and moods. We all share the same elements crafted by early evolution: pain, pleasure, fear, hunger, bonding, cyclical conscious and unconscious states. We imagine other animals as peasants and ourselves as lords, but we share a lucky heritage and a lavish kingdom.

CHAPTER 33

A Kingdom of Neighbors

"A brain is a very mediocre commodity. Why, every pusil-
lanimous creature that crawls on this earth or slinks
through slimy seas has a brain!"

— Great Oz to the Scarecrow,
The Wizard of Oz
MGM Studios, 1939

The hummingbirds depart as summer wanes, which may be why
it feels like such a loss. I'll miss their iridescent flurry each day,
and how at night they darken and linger, sip seven or ten times in
a row, hovering backward after each sip to lift swordlike beak, slowly
swallow the liqueur, then belly up to the feeder again for a last of last
nightcaps. I'll miss these small emissaries from the natural world
who share our bounty and fill our lives with jeweled eyefuls each
summer. Last year the adults left suddenly on September 5, as if
they had a train to catch. This year they already seem restless and
may leave earlier. A hormone is quietly tugging at their brains, making them feel jumpy, ready to ramble, and desperately hungry.

I'm sure they're not aware of time, don't know where they're
headed, can't picture jungle flowers oozing nectar, won't remember
much. Yet they navigate better than we do. Bird brains evolve
extra neurons, get rid of those they don't need, and grow more that
they do. A child's brain behaves much the same. Because it's always
easier to work with too much than too little, a lot of leafing out and
pruning goes into sculpting a brain, whether it's a hummingbird's

or a child's, and, astonishingly, each brain knows what topiary to shape.

But unlike us, birds don't need to plumb the past to solve present problems. Like all animals, they sense and feel, but most don't need to think about it, because body wisdom guides them with instincts and reflexes. A squirrel needn't remember where it buried acorns three years ago, only where it buried fifty this year. Anyway, instinct leads it to likely spots. Each species evolved a brain according to its special needs. But are they conscious?

How about the brain of our nearest relative, the chimpanzee? Chimpanzees feel deeply, strategize, plan, think abstractly to a surprising degree, mourn, experience pleasure, empathize some, deceive, seduce, laugh, and are all too conscious of life's pressures, if not its chastening illusions. They're blessed and burdened, as we are, by strong family ties and quirky personalities, from madcap to martinet. They jubilate when happy, mope when sad. I don't think they fret and reason endlessly about mental states, as we do. They simply dream a different dream, probably much like the one we used to dream, before we crocheted into our neural circuitry the ability to think about thinking, know we know, know what we don't know, guess what others are thinking, project our feelings onto others, assume others share our opinions, and talk about notions and behaviors. One result of this may be our yen to anthropomorphize everything from dolls to dogs. We're such lonely beings that we sometimes attribute our exact minds to other animals, especially pets. Because we can project ourselves into the minds of other people, we tend to endow everything with a human mind—dolls, ocean, sun, wind, other animals, plants, volcanoes, statues, landforms. Then we can attribute to them all the things about ourselves we can't stand.

Perhaps we're closer cousins than we'd like to believe? We share much of our emotional and intellectual heritage with chimpanzees and roughly 95 percent of our genes. Including a language gene, FOXP2, which is quiet in chimpanzees, very rowdy in humans. That means that about 200,000 years ago, we shared a common ancestor who probably grunted, whistled, and made other sociable noises. In us, a variant of FOXP2 evolved, refining speech, and proving such a boost that it quickly percolated through our population. One study of chimpanzees and humans revealed that a few of the human subjects were genetically closer to chimp than to human! (When I mentioned this to a girlfriend, she swore she had dated one of them.)

In my mind's eye, I see a human face and a chimpanzee face side by side. How can we be so similar genetically and yet so different psychically? Svante Paabo, of the Max Planck Institute in Leipzig, compared the brain activity of humans and chimps. Our brain is twice the size of a chimpanzee's, but, except for size, the basic anatomy looks the same. So do many of our social skills, from preserving family ties to kissing up to higher-ranking people. Using a tool known as a gene expression chip, Paabo found striking differences in how chimpanzee and human genes behave. Although human and chimpanzee genes behaved similarly in liver and blood samples, they varied dramatically in the brain, where human genes authorize the manufacture of many more proteins. Because of that, our neurons sizzle with activity. How that translates into differences in mental gait, into our slabs of wistfulness, or into lost love that feels like phantom limb pain, is anyone's guess. But it confirms what we already knew, that we angled off from our close cousins 6 or 7 million years ago, and developed our own style of mental mischief and mayhem. That includes daring language skills, as well as a prolonged childhood—almost a third of our lives—and delayed sexual maturity. Those three features alone have whittled and crafted us in countless ways. But the more we learn about chimps and bonobos (pygmy chimps) and orangutans, the more we discover family traits.

Look in a mirror and an alien face may stare back at you, maybe the face of your mother or father, or a suddenly wrinkled or sunburned face. Unnerving as that spectacle can be, you don't search behind the mirror for the person. Except for an unlucky few people with brain damage, you know it's the one and only you, reflected in a frozen pool of glass. A *reflection*: a sight so powerfully evocative that we turned it into a metaphor for an important habit of the mind. *Know thyself* hinges on the same ambiguity. We find that ancient, mythic test of self-awareness so commonplace, and yet few animals can pass it, including human babies for their first two years or so. The mirror test has become the gold standard for judging a creature's self-awareness. Chimps recognize themselves in a mirror, and there are reports of dolphins doing the same.

Killer whales have evolved different cultural traditions for hunting, communicating, and socializing. Borneo orangutans make flyswatters, use leaves as gloves or napkins, play competitive games, use objects as sex toys, and bend their hands or leaves to control sounds—behaviors that differ among groups, and are clearly learned. Some other animals have evolved basic cultures.

Surely abstract thought is the high bar, and no other animals could mull over a phrase like *a hundred thank-you's in a shoe of shame*? Probably not. But many other animals can reason abstractly. We're not rare because we have wildly different minds from other animals. Our minds just do *more* of some things: more self-awareness, more reasoning, more speech, more worrying, more abstract thought, more inventing, more calculating, more analyzing, more empathizing, more problem solving, more feeling. The payoff is our symbolic, narrative sense of self. Our unique niche is to be more nimble-minded than other animals, but the world is storied with clever, feeling animals.

Why did we diverge and follow our own genetic path? The seeds of change may have been viral. Geneticists have found that nearly half our genes are "junk DNA" possibly installed by viruses millions of years ago. The molecular evolutionist John McDonald, of the University of Georgia, and NIH's King Jordan studied a family of 147 transposons (viral DNA that can move around our chromosomes) in various primates. They found the transposons in humans, but not in chimps. Humans and chimps may have diverged because viral bits of DNA invaded our chromosomes, tinkering with some genes and the proteins they produce. Or they may have wreaked havoc, prompted widespread changes, and revised our entire genome. "We like to think that our DNA must be serving us," McDonald observes, "but the vast majority of our genome is not directly related to our own function. We're just part of a larger pic-ture." A picture of body snatchers, which inserted their DNA sequences, redirecting our evolution in grand and subtle ways. We may have parted company with chimpanzees millions of years ago, but there's nothing purebred about us. We're mongrel com-mittees of cells, the patchwork destiny of different life-forms.

It's also possible the junk DNA isn't really junk at all, but little-known gene switches, silencers, and regulators; pseudogenes, and other meddlesome stuff. In time, we're bound to discover how the "dark parts of the genome" enrich and bedevil us, and help make us unique.

We do share some of our motives, feelings, and instincts with other animals, and pretending that we don't makes no sense. We share fear, curiosity, sexual desire, hunger, social contact, defen-siveness, hoarding and saving behavior, status seeking, dominance, pleasure, attachment and protectiveness, avoidance of pain and danger, grooming behavior, play, aggression, exercise, acceptance, appeasement and submissive behaviors, family loyalty, and a desire for restful calm. What we honor as a search for truth and knowl-edge, other animals may experience as exploratory curiosity. What we call freedom, other animals might experience as a powerful instinct to leave home in search of resources, stimuli, and a mate. Many animals (and some plants) also communicate in basic ways, including musically.

Male Borneo tree frogs use trees as musical instruments. They select a tree with a hole partly filled with water, half submerge in the puddle, and then chirp moonlight come-hithers to females. At the start of an evening's concert, a frog will raise and lower pitch, until he finds one that resounds, then he'll lengthen and shorten his call until he gets into his groove, performing loud and ardently, chirping his heart out. Testing notes until he finds the right one, analyzing the acoustic feedback, fiddling around with the phrasing until it sounds perfect—that sounds like music making to me. Also to Bjorn Lardner, of the University of Lund in Sweden, who stud-ies tree-hole frogs, sometimes giving them artificial wind instru-ments such as a pipe with a niche partly filled with water. As a male starts his chirping warm-up, Lardner changes the water level, and sure enough, the frog adjusts its chirps to the changing pitch of the pipe, until it finds the one that resounds. Following Lardner's advice, this morning I imitated the vibrant Borneo tree frogs by humming in the shower and changing pitch until I found one that made the stall resound.

Then there are the amphibious lounge lizards, the jazzy *Amolops tormotus* frogs, in Anhui Province in China, that warble tunefully in the underbrush and ad lib countless ditties. When researchers analyzed twelve hours of taped singers, they were surprised to discover that no two males ever sang the same song, and no frog repeated any of his own songs. Richly varied, with many vocal swoons and fugues, their nimble stylings sound much like the songs of birds, whales, and primates.

But music isn't enough. Some people argue that to be conscious an animal must speak a true language and understand its symbols to several degrees—for example, thinking about how the symbolism of thinking about symbols influences further thinking about those symbols and how that may alter neural networks in such a way as to favor such thinking in the future. Any animal that can't reason isn't conscious, they conclude. But most of our own thoughts aren't conscious, and yet we make choices, act purposefully, engage in behavior as complex as driving a car without actively thinking about it.

Others claim we're the only animals that have a subjective, phenomenal experience of the world. Anyone who has closely observed animals in their natural habitat knows that more-sensitive-than-thou chauvinism to be wrong. I'm sure squirrels have a subjective phenomenal experience of the world because I've spent a lot of time observing their habits and senses.* Unlike most wild animals, squirrels will look right at you and hold your gaze. (Well, I say *at* you, but their bulging eyes can see sideways at the same time.) When they chew, their mobile cheeks move a lot and their long

*Which I studied for *National Geographic* and chronicled at length in *A Slender Thread*.

whiskers twitch. Whiskers are sensing organs, and squirrels feel the air, snow, wind, rain as they eat, which may enrich the pleasure.

One day, I saw a dominant male squirrel, who had been ousted by juveniles, drag in after most of the other squirrels had fed. He moved tentatively and didn't approach the house—perhaps because he had a better chance to escape in the open. He ate a few nuts, slowly, in a kind of trance, while other squirrels growled, and then one attacked him and he leapt onto a hickory sapling. A hairless patch on his back looked larger. Had he been assaulted again overnight? His personality change was startling. For a while, there weren't enough nuts in the world to eat or store. He wanted all of them, and battled any squirrel that inched near. Now he seemed lethargic, scared, and watchful. Moving to the edge of the group, he climbed a tree and sprawled listlessly on a limb, descending to feed only when the others left. A human in that condition we would call depressed. I was doing telephone crisis counseling during the same years I studied squirrels, and I noticed that the squirrels and humans suffered from similar distresses: lack of resources, lack of attachments, lack of status. There were countless subtle similarities. For example, male squirrels would often hang out with female familiars. When breeding season came the females would allow those males to catch them and breed. Or, if you prefer, the females favored friends, a behavior the males knew and exploited. Did they think about it ahead of time, when they befriended the females? Not the way we would. But I'm not convinced our style of reasoning is the only form of so-called consciousness. If we're just talking about self-awareness, social nous, mood changes, distinctive personalities, and having subjective phenomenal experiences, then many animals qualify as conscious.

What should we make of Fu Manchu, the male orangutan that kept escaping from his pen at a large city zoo? Understandably bored and curious, he would roam the zoo and watch the herds of humans. How did he escape time after time? Using a video camera, the staff finally discovered that the orangutan would hide an unfolded paper clip beneath his upper lip when people were around. If he thought the coast was clear, he would bend the wire into the right shape, pick the lock, let himself out of the enclosure, then straighten the paper clip again and hide it under his lip for future use. Doesn't that sound like the sort of jailbreak we might expect of a bored, stir-crazy human? Among other things, the orangutan had to be aware of how his body shape appeared to humans—that if he hid the paper clip inside his lip it wouldn't be visible. He had to be able to hold the thought of someone thinking about him. He had to know that he was doing something illicit. That's a lot of mentalization for a supposedly nonconscious animal.

We used to single ourselves out as toolmakers, but as I said earlier, chimpanzees, bonobos, and orangutans have all been observed using tools. And tool use isn't limited to primates. In the August 9, 2002, issue of *Science,* for example, researchers tell of tool-using New Caledonian crows, one of whom really surprised them when she figured out that by bending a stray wire into a hook, she could fish a bucket (containing food) out of a pipe. Neuroscientist Jennifer Mather, of the University of Lethbridge in Canada, has documented octopuses that use tools and spend time playing. Many animals have discovered medicine. Some of my favorites are the capuchin monkeys that rub themselves with a millipede as an insecticide. They tend to lather on some millipede juice and then pass the bug along to a neighbor as if it were a bottle of Cutter's. Entomologist and chemical-ecologist Thomas Eisner, of Cornell University, discovered that the millipedes secrete benzoquinone, which works as a cleanser and an insecticide. So many wild animals use medicines that there's a field of study devoted to it: zoopharmacognosy.

How about a feeling we take pride in, such as empathy? I'll never forget sitting quietly among the leaf litter in the Brazilian rain forest, and watching the soap-opera-like dramas of golden lion tamarins, beautiful kabuki-faced monkeys that are vanishingly rare. Tamarins are monkeys driven hard by instinct, and they don't need higher thought to conduct complex social relationships. In one tamarin family, the mother and adolescent daughter fought viciously over a new male and the loyalty of the younger siblings. The distance between the tamarin brain and ours is vast, yet they behave in some recognizably human ways. They act as if they can anticipate each other's actions, predict cause and effect, intuit motives. Does empathy require our kind of thought, our kind of consciousness?

Many mammals experience empathy to some degree, and it makes sense that particularly social animals would. I think psychologist Michael Lyvers, of Bond University in Australia, makes an excellent case for its usefulness. As he writes in *Psyche*, social animals use empathy to great evolutionary advantage. It helps social animals read each other and figure out how to act.

> A being who lacks empathy . . . would more often tend to respond inappropriately to others in their social group, and would probably find it quite natural to imagine that others lack a subjective or "feeling" experience of the world (perhaps this is how psychopaths perceive other humans and non-human animals).

In its simplest form, empathy offers "monkey see, monkey do" learning. Special mirror neurons help in the pantomime, and some of the empathy is observable through brain imaging. A monkey can watch another monkey toss a ball, and the same parts of his brain will glow as when he himself tosses a ball. We've probably just elaborated on that evolutionary gift, as we have on so many others.

Does any other animal share our mental fantasia, our sort of consciousness? By definition, they can't. It's a term we invented to apply to us alone, since we're the only consciousness we really know. "Just thee and me," as the saying goes, "and I'm not so sure about thee." After all, it's hard to believe one's mental circus is shared even by other humans, the acrobatic plays of thought seem so personal, peculiar, one of a kind.

CHAPTER 34

The Beautiful Captive

We die containing a richness of lovers and tribes, tastes we
have swallowed, bodies we have plunged into and swum
up as if rivers of wisdom, characters we have climbed into
as if trees, fears we have hidden in as if caves. I wish for all
this to be marked on my body when I am dead. I believe
in such cartography—to be marked by nature, not just to
label ourselves on a map like the names of rich men and
women on buildings. We are communal histories, communal books. We are not owned or monogamous in our
taste or experience. All I desired was to walk upon such an
earth that had no maps.

—Michael Ondaatje,
The English Patient

The brain is still terra incognita on the map of mortality, still the
fabled world where riches and monsters lurk. But we've begun
mapping its shores and learning about its ecology. Bridging the cre-
vasse between curiosity and necessity, we've devised ways to look
within ourselves without opening the skull. The most recent fMRI
photographs are beautiful as abstract art, an exciting panorama of
shapes and colors; and they are beautiful as an idea: the mind made
visible. When you realize that you are looking at the partial snapshot of a mood, and pause a moment to appreciate the fullness of all
that implies, then beauty looms in several dimensions. So that's a
shard of pain, you might find yourself saying, so that's what jealousy
looks like. Isn't every piece of music a snapshot of a mood? An

entire novel might well be required for the event being remembered during one brain scan. A duet of mystery and suggestiveness adds to the images' beauty, as it does to the moving mosaic of the mind.

Like impish spirits surveying from above, we may lift the lid off the brain with PET, MRI, fMRI, or MEG and peer inside, but that still leaves us voyeurs, distant viewers. Our machines can only tell us so much. They're good diagnostic scouts, when we're hunting a tumor or trailing a stroke. Used experimentally, to understand how the brain works, they're more fallible and easily biased. Our new mechanical eyes, replacing the past's newest eyes (magnifying glass, X ray, and so on), will join the heap of outmoded technology one day, replaced by a machine that reveals the brain's full star chamber in time, microsecond to microsecond, as it surges with electricity and its chemical stew simmers. At the moment, we're a little like lepidopterists studying a butterfly pinned in a velvet box—a beautiful and instructive icon, but one sprung loose from its processes.

Throughout this book I cite brain imaging studies, which tend to be more suggestive than absolute, since the beautiful captive inside the skull often eludes our lumbering machines. We try to map it in revealing and useful ways as it pictures an apple or face, reads a word, laughs, smells, reasons, anticipates pain, wrestles with moral dilemmas, gets ready to move a toe, feels lusty, is hyperactive, reveals the fingerprints of a disease. We measure its blood flow, hunger for glucose and oxygen, its radioactivity or magnetism. Regions light up and offer us some illumination. But the porch light can be on without telling you who is at home. When it comes to brain mapping, "our knowledge is akin to looking out of the window of an airplane," neuroscientist William Newsome, of Stanford University, explains. "We can see patches of light from cities and towns scattered across the landscape. We know that roads, railways, and telephone wires connect those cities, but we gain little sense of the social, political, and economic interactions within and between cities that define a functioning society." I like that analogy. Also, we should remember that detecting a hole in the ozone layer over Antarctica doesn't mean that Antarctica's frozen deserts caused it. There's no hint that the hole arose from a worldwide network of events in a dynamic, constantly changing social climate and fluid weatherworks.

So the brain maps aren't perfect, but they're nonetheless producing wondrous insights about the brain, especially when techniques are combined, since each machine has different strengths and weaknesses. PET (positron-emission tomography) produces three-dimensional images of blood flow in the brain by using radioactive particles; MRI (magnetic resonance imaging) creates even more detailed images by bombarding the body with a steady magnetic field and radio pulses; fMRI (functional magnetic resonance imaging) creates images by measuring oxygen levels in active neurons; and MEG (magnetoencephalography) picks up the weak magnetic fields broadcast by neurons as they change from moment to moment. Computers paint their results by number. In all cases, the colors are arbitrary, a human device to clarify differences in temperature. The same is true of the sumptuous colors of objects viewed by telescopes—to the naked eye they look black, white, and gray. Scientists colorize computer data to better suit our senses, which relish color, and also rely on color cues to divine depth and distance. How curious that we instinctively identify heat with red and yellow, and cold with blue and green.

Since beauty is a phantom of the mind, it's fitting that we find the images sensuously beautiful, even if we don't need to. The sound of cicadas on a sultry summer evening didn't evolve to please us. Attuned to ultrasound, cicadas don't even hear what we hear, yet their voices delight us, just as color snapshots of the living brain do.

Among the many exciting truths offered by brain imaging is that talk therapy and drugs can be equally powerful treatments. Change a distorted way of thinking and you change the brain. Change the brain and you can change some distorted ways of thinking. Not all. Schizophrenia and other mayhems resist. But obsessive-compulsive disorder, for instance—which may stem from a brain region vital to filtering out mischievous thoughts and impulsives—responds well to cognitive behavioral therapy. Psychotherapy's insights rely on deep learning, and learning of any kind alters the brain. Brain imaging confirms this with vital snapshots of the process. As Eric

Kandel has shown in classic work on the nervous systems of sea slugs, when any animal learns it produces new proteins, which in turn create more connections among the neurons. Talking with someone can sway brain function. For many people, taking anti-depressants helps to install new memories at precisely the time talk therapy is providing important things to remember. Brain imaging, used to show the work in progress, can help decide a treatment's efficacy. PET scan studies have found that as patients improve, their brains show the same changes whether they received antidepressants or psychotherapy.

If we can't view the brain's galleries as intimately as we'd like, at least we can glimpse them with machines created by inquisitive minds. I like the way Kathleen O'Craven of Toronto's Baycrest Centre for Geriatric Care and Nancy Kanwisher of MIT want to use brain scans. Able to tell if someone is imagining a place or face about 85 percent of the time, they're hoping to enter the clenched mental world of stroke victims who can't talk.

What a boon brain imaging is as a diagnostic aid in medicine and surgery, and also as a suggestive guide to the brain's intricate wilderness. I find this a little like looking at images of Earth from space. When we stare down at the ocean we can learn much from the patterns and colors and prowling weather systems. Yet a scaly universe lies hidden beneath the waves.

A line of imaging I find especially intriguing explores murderous brains. If we peered into a serial killer's brain, would we see niches filled with the equivalent of a tiny skull and crossbones? Or, much more worrisome, would it look the same as everyone else's? Psychologist Adrian Raine, of UCLA, studies the biological basis of criminal behavior by imaging the brains of convicted killers. A PET scan, spying on a brain from above, vividly displays the metabolic activity in patches of red and yellow (very active) or blue and green (not so active). "What we forget," Raine explains, is "that for 99 percent of their lives, murderers are just like us. Tragic actions in the other 1 percent of their lives set them apart . . . the structure and functioning of their brains may also set them apart." He believes he's found "the first evidence . . . that the brains of a large sample of

murderers are functionally different from those of normal people."
The results of his studies are provocative: murderers seem to have
less active left hemispheres, more active right hemispheres. Their
PET scan "signature" involves the prefrontal cortex, amygdala,
corpus callosum, and hippocampus. What's more, Raine found distinctive brain differences between predatory and emotional killers,
and between killers from "good" homes vs. "bad" homes. Might
there be a social basis for the biology? Here's one likely scenario
apropos of the corpus callosum, that busy highway of fibers connecting the brain's two hemispheres. An abusive parent who shakes
an infant might weaken the fibers, and when the two sides of the
brain can't communicate well, the right hemisphere's negative
emotions are unbridled by the left. Add that to other brain abnormalities (impelling aggression, poor reasoning, recklessness, lack of
self-control, and mental inflexibility, among others), and the result
can be lethal. As Raine rightly warns, these revelations may inspire
feverish moral, political, and legal dilemmas, such as this one: "If
brain deficits make a person more likely to commit violent crime,
and if the cause of the brain deficits was not under that person's control, then should he be held fully responsible for his crimes?" Suppose a future husband, student, or employee were asked to produce
a brain scan and the results showed similar abnormalities, despite
testimonials to his peacefulness? Without a library of brain maps,
we don't really know what's normal (a kingdom that usually
includes a wide frontier), or how often people with a killing-sig-
nature brain may be pacifists. Playing devil's advocate, one might
argue that in addition to the problem of "normal" abnormalities
(though dyslexics also tend to have less active left hemispheres, few
of them become violent), there is the uncomfortable truth Raine
mentioned—that most of the violent among us are neurologically
normal. In any case, compiling a large atlas of brain maps, the mental equivalent of fingerprints, might expose people to all manner of
policing or prejudice. Brain imaging can't solve every riddle, and it's
not flawless, or without ethical mazes, but it has already provided
a rich harvest, and a gallery of beautiful vistas, as it continues to
explore the mountains of the mind.

Although brain research is in a great era of discovery, much still lies hidden beneath the waves. But we are the animal who questions, who has invented ways to extend its senses. We are the animal that studies itself. The animal that worries. The animal that lies most easily and most often. Somewhere in the human beast, dreams are made, art is created, romance unfurls. Not in especially simple or efficient ways. On average, we spend two weeks of our lifetime waiting for a traffic light to change. We stain everything with feelings. We are the animal that elaborates. We who invent bubble-gum-scented pens, tongue piercing, blue roses, orange-starfruit-chamomile tea, and other sensations in which to bathe. We who add to life such spectacles as the "all male chain-saw marching band," a regular feature of the Ithaca Festival's annual parade (as is the dancing "Volvo ballet corps").

We who adopt such job titles as:

ramrod, doper, animal impersonator, aquarium tank attendant, colors custodian, alligator-shear operator, egg smeller, hooker, impregnator, dog-food-dough mixer, matzoh-forming machine operator, antisqueak applier, cookie breaker, dead header, fish flipper, pantyhose crotch-closing machine operator, maturity checker, sea-foam-kiss maker, lingo cleaner, bosom presser, banana-ripening-room supervisor, automatic lump-making machine tender, jollier, rump sawyer, Panama hat smearer, puddler, gang vibrator operator, clam sorter, fig washer, upsetter, armhole feller, brain picker, leaf suck operator, hand pouncer, fur beater, necker, mangler, belly wringer, cat chaser, head chiseler, jawbone breaker, skunkskin curer, crayon grader, specimen boss, apron scratcher, wet-end tester, napkin-ring wrapper, guillotine operator, roughneck, suppository molder, wrong-address clerk, bonbon cream warmer, king maker, Easter bunny, kiss mixer, pillowcase turner, and mother repairer.*

*All are jobs currently being held by American workers, according to the U.S. Department of Labor, which includes about 25,000 more in its *Dictionary of Occupational Titles.*

We who indulge in the naive belief that we can re-create the whole world, even though, as the painter David Hockney points out, "Well, you can't. Where would you put it? Next to the whole world?" We, who strive to understand ourselves, accept ourselves, outlive ourselves. Why we elaborate is itself an interesting question. After all, procreation doesn't require taffeta prom dresses, survival doesn't depend on pantyhose, racquetball, or the giant naked Hercules sculpted out of chrome car bumpers that lords over a quad at Cornell (every spring students dress his penis in a colorful condom). Do we elaborate to such a degree because our brain is spinning its wheels in a world where for the most part it's not required to make life-and-death decisions each day? Do we elaborate because our evolution, attuned to simpler times, made sure that the sheer ability to elaborate *anything* gives us a special piquant pleasure? Do we elaborate to strengthen our illusory sense of control in a random universe, where we're prey to forces beyond our knowledge or influence, and from which we will ultimately die? Do we elaborate as a form of breeding competition—because a nimble-minded mate will have a better chance of surviving and rearing offspring? Do we elaborate because concocting variations upon a theme is simply a basic force in nature, whose habit is to see how many ways something can be used (consider eyes)? Do we elaborate as a form of trading resources? Do we elaborate as a way to limit and separate, accentuating the clear distinctions between things? Do we elaborate to occupy a larger space in life, as a superstitious ploy to subdue death? Do we elaborate to soothe ourselves when life feels too abrasive, the way an oyster exudes nacre to coat irritating grit? Do we elaborate just for fun, as a kind of sensory sport? This list of possibilities could sprawl for a while longer, since none of them is provable, and they may all be right to a degree. Elaborating on why we elaborate is its own reward.

Our brain is sloppy and imprecise, but that's its strength. We strive to be orderly, evolution doesn't. It adds on, tinkers, reuses parts. Evolution favors anything that works, no matter how wacky. It chooses easy over best, quick over precise. This doesn't result in perfect designs, but in good enough. When it comes to creative solutions, messy offers far more scope than tidy, and gadgets prosper where precision instruments fail. One year, as an adult, I asked for a little boy's Christmas and received a chemistry lab and three different erector sets to play with—one American, one British, one German. The American set included unpainted metal parts you could fix together in countless sloppy ways. The British set was a little more refined and included some colorful parts. The German set was beautifully designed, the parts exquisitely engineered, but at a cost. With that much precision, you needed a plan before you began building.

It sounds clinically cold, maybe profane, to refer to our brain-work—so imaginative, thoughtful, insightful, logical, and creative—as a kind of *information processing*, a term most often applied to the handiwork of computers. We've fashioned computer brains with the same basic on/off technology of our neurons. But we create based on what we know, even if it's a limited facsimile. The brain's simultaneity, parallel processing, subjective feedback, illogic, emotional cauldron, and, at times, sheer gratuitousness aren't things a computer, however sophisticated, could double. Computers master some things better than a brain can, most things worse. We're enlarged by the talents of our creations, not reduced by their limits. As I suggested earlier when praising the respectability of matter, maybe information processing isn't as mere as we suppose, maybe it's the soul of inquiry, and can be performed superbly in limited, rudimentary ways even by the nonliving systems we lively systems create. Living systems like brains do better with the untidy, inexact, but versatile approach. It's jumpier but more flexible. Although most brains learn well, and are superb at finding patterns, they can't compete with a computer or even a calculator for basic math, and they're tragic at logic, which most of us learn the hard way, through trial and error, if we survive the first thousand or so humiliations. Unlike precision instruments, brains don't need to be accurate all the time. Good enough often enough will do, even if that means skimping, doubling up, or sheer gamble. A limber brain is a successful brain, however sloppy. A precision brain is a computer. A perfect brain doesn't exist.

Gerard Manley Hopkins was right, the brain has vertiginous .
mind cliffs "no-man-fathomed." Wallace Stevens was right, the
body can be satisfied often and well, but the mind never. We
move from tigerish to tender, watching snowflakes fall like shred-
ded doilies. We hear rat-pup laughter in the rain. "The mind is its
own beautiful prisoner," e. e. cummings wrote from one of love's
many penitentiaries. Because mind is matter, we matter and we
mind. We use words like small hand axes. We televise alternate
worlds. We scout the invisible. We practice the art of science. As
songsters and sages have always said, we dream several dreams, only
one of them during the night. For better or worse, we've become
nature's way of thinking about itself, a brain for all seasons.

ALCHEMICAL SYMBOLS

 TORREFACTION OF SILVER

 TO MIX

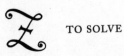 TO SOLVE

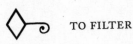 TO FILTER

 ESSENCE

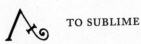 TO SUBLIME

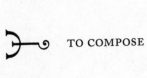 TO COMPOSE

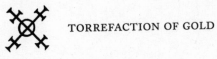 TORREFACTION OF GOLD

NOTES, ADDENDA,
AND AFTERTHOUGHTS

1. The Enchanted Loom

The title of Ravel's beautiful pavane, "Pour une Infante Défunte," is often mistranslated as "Pavane for a Dead Infant." A pavane is a dance and he first wrote this one for piano, in 1899, dedicating it to the Princess de Polignac. Most conductors render it as a melancholy lament for a dead child, and it's exquisitely soulful in that form, but Ravel composed it to be played at a faster tempo, as a dance. In 1910, he orchestrated it, and most people find the orchestral version a wonderful elaboration. In a 1912 interview, Ravel was critical of his extremely popular pavane, attributing its success to "the remarkable interpretations" inspired by its title.

2. This Island Earth

"O brave new world,/that has such people in't." William Shakespeare, *The Tempest*, V, i, 183.

3. Why We Ask "Why?"

Even the neurons in the right and left hemispheres vary, which may mean that they form memories differently. Isao Ito and his team, at Kyushu University in Fukoka, Japan, have found more of a specific type of NMDA receptor (important for learning and memory) on dendrites at the *tip* of neurons in the right hemispheres of mice, but at the *base* of neurons in the left. Reported in *Science*, vol. 3000, p. 990.

William Gass, *Tests of Time* (New York: Knopf, 2002), p. 27.

Michael S. Gazzaniga, "The Split Brain Revisited," *Scientific American*, August 31, 2002, pp. 27–31.

Our brainpower doesn't come solely from having left-right brain differences, a design feature we share with the higher primates and some other animals.

4. The Fibs of Being

In famous experiments, the neuroscientist Benjamin Libet of the University of California at San Francisco tested people undergoing brain surgery for epilepsy. Since brain cells feel no pain, and the patients had to be awake anyway, they were able to respond to Libet's probing questions and fingers. When he touched a patient's hand, it took about half a second to become *felt*. If, instead, he touched the gray matter itself, right where the somatosensory cortex processes stimuli from the hand, it still took half a second. The brain needs time to circulate its news.

In a related experiment, Libet capped students with electrodes and asked them to move one hand whenever they felt like it, and note the precise time of their decision. An EEG revealed brain-wave activity *before* the students decided to act. Libet repeated the test hundreds of times, with the same startling results. EEGs kept showing the cerebral cortex becoming active, awareness of an urge or decision to act about half a second later, and finally the hand moving two tenths of a second after that. Shouldn't effect follow cause? Since a tingle or tiding can take half a second or so to hit consciousness, the mind *pretends* it's half a second earlier, so that the two will feel simultaneous.

Libet believes that consciousness evolved to monitor the brain and intercept its choices when need be, during that two tenths of a second between urge and action. It would have to do that without being preempted by the brain.

Michael S. Gazzaniga, "The Interpreter Within: The Glue of Conscious Experience," *Cerebrum* 1, no. 1 (spring 1999): 68–78.

Self-awareness and working memory; a woman berrying. Telephone conversation with Larry Squire.

Kevin N. Laland of the University of Cambridge in England. Bruce Bower, "Evolutionary Upstarts," *Science News* 162 (September 21, 2002): 186.

See William James's essay "The Dilemma of Determinism," in *William James: Writings 1902–1910* (New York: Library of America, 1988), pp. 536–38.

5. Light Breaks Where No Sun Shines

Sigmund Freud, *The Standard Edition of the Complete Psychological Works of Sigmund Freud,* ed. and trans. James Strachey (London: Hogarth Press, 1953–74).

Guy Claxton, *Hare Brain, Tortoise Mind* (Hopewell, N.J.: Ecco Press, 1997), p. 49.

Amy Lowell, quoted in *Hare Brain, Tortoise Mind,* p. 60.

Alfred North Whitehead, quoted in *Hare Brain, Tortoise Mind,* p. 15.

6. The Shape of Thought

Francis Crick, *The Astonishing Hypothesis* (New York: Simon & Schuster, 1994), p. 103.

Here's a humbling thought: we possess only about 35,000 genes, and we share many of those with mice and insects. A worm has 18,000 genes, a fly 13,000. Some plants have about as many genes as we do. If genes contain DNA—the blueprints for building a wombat's or a child's brain—how can so few genes result in the complicated beings we are? Life is agile. It's not the size but the creativity of the genome that matters. Take a limited number of genes, change their how, when, where, and how long, and you get radically different anatomies and behaviors. Some genes are identical in a wide variety of animals, but molecular coaches force them to express themselves differently. Because nature loves economy, genes can be as versatile as a child's building blocks. Using identical nails and struts, it's possible to build a wing, a boat, a Ping-Pong table.

We used to think of genes only in terms of heredity, as old news, a spiral scroll of blueprints, without appreciating their role in our minute-to-minute dramas and emotions. Everything about us, from hallucination to rectum, every *eureka!* and descant, owes some debt to genes. It may be as iffy as a predisposition or as concrete as a sixth finger. Individual genes don't correspond to individual connections, most of which are forged on the anvil of life.

Imperious genes order the making of proteins, whose shape dictates how they'll behave in the body, so they must fold correctly (misfolded proteins are thought to contribute to Alzheimer's and other illnesses). Like a falling raindrop, a protein changes shape when it encounters others. As its shape shifts, so does its purpose, which is why animals with similar genes can look very different.

There may well be a million proteins, each one cooked up by amino acids. How beautiful amino acids look through a scanning electron microscope, lit by polarized light: pastel crystals of pyramidal calm, tiny tents along life's midway. To our gaze they seem gemlike and arid. We cannot see their vitality, how they collide and collude as they build behavior. A guidebook would help. No one has created a human protein map yet, nor even fully understood proteins. What we really need is a mobile map, not a list of statuary, to discover a protein's destiny. Because we're voyagers by design, we're bound to sketch one soon. We already map mountains, continents, ocean bottoms, the trickling source of great rivers, cities, organs, and moon maria. We've charted our genes. An atlas of proteins will greatly add to our library.

Work by Svaboda and Gan, Jaime Grutzendler, et al. "Long-Term Dendritic Spine Stability in the Adult Cortex," *Nature* 420 (December 19/26, 2002): 812–16.

In research with rats at the University of Illinois, animals raised with playmates, toys, and other sensory stimuli, in an enriched environment, grew 25 percent more synapses than other rats, and also grew more neurons in a memory area of the brain. "Gray Matters: The Arts and the Brain," a radio program produced by the Dana Alliance for Brain Initiatives and National Public Radio, September 2002.

If he devoted only a second to each connection, a final sum would take 32 million years. See Floyd E. Bloom, et al., *The Dana Guide to Brain Health* (New York: Free Press, 2003), p. 11.

Concerning inner electricity: "The neuron has a resting potential across its membrane. This is typically about -70 millivolts (inside versus outside). A change that makes this more positive at the cell body (say, -50 mV) is likely to make the cell fire. One that makes the potential even more negative can prevent it from firing at all." Crick, *Hypothesis*, pp. 97–98.

Salvador Dalí would have liked knowing about the bending columns in the brain, vertical strips of cells that work together. They are forwarded news from elsewhere and cooperate on different jobs. One column, the sensory cortex, handles the body's physical sensations. Another column, the motor cortex, cranks up voluntary movements. Busy body parts like the tongue, feet, and fingers demand more space. Are the columns Ionic, Doric, or Corinthian? They're Dalíc, with uneven edges of living matter, and supple folds tinged pink with blood.

7. Inner Space

E. M. Cioran, *A Short History of Decay*, trans. Richard Howard (New York: Viking, 1975), p. 105.

Antidepressants may also spur new cell growth in the hippocampus, a key region for memory and learning. That would explain why it takes them about a month to lighten one's mood, the same time required for cells to develop. See René Hen, *Science* 301 (2003): 805. Also Fred H. Gage, "Brain, Repair Yourself," *Scientific American*, September 2003, pp. 47–53.

Crick, *Hypothesis*, p. 104.

Floyd E. Bloom, et al., *The Dana Guide to Brain Health* (New York: Free Press, 2003).

8. Attention Please

Crick, *Hypothesis*, p. 92.

The multitasking MRI study was led by the psychologist Marcel Just of the Center for Cognitive Brain Imaging at Carnegie Mellon University, and reported in *NeuroImage*, August 1, 2002.

I met a young woman who shared a story with me about how well animals pay attention. "When I was six," she explained, "one side of my extended family gathered at my parents' house for Christmas. Usually my parents, my sister, and I shared the holidays with my grandparents, but this year an aunt and uncle flew up from Florida and another uncle flew in from Texas. Harry, our two-year-old cat, who walked me to the school bus stop every day and waited for me to board the bus before he left for his own adventures, wasn't used to so many people, so he escaped outside on Christmas morning and watched us for a long while through the window as we exchanged gifts. Eventually, he left his vigil and stalked off. Not too long after that, a scratching at the front door let us know that he wanted to come in. As soon as my dad opened the door, Harry trotted across the room, straight to where I sat on the floor opening presents. He was carrying a present of his own, an offering: a small bird, just for me. I was thrilled and honored."

Hegel quoted in Donnel B. Stern, *Unformulated Experience: From Dissociation to Imagination in Psychoanalysis* (Hillsdale, N.J.: The Analytic Press, 1997), p. 63.

At Johns Hopkins, some researchers have been studying how neurons pay attention. Monkeys were asked to perform two different visual tasks. Whenever they switched from one to the other, a "chorus" of neurons fired together in the relevant area of the brain, perhaps to create what we call *focus,* or maybe to outshout other sensory distractions.

10. In the Church of the Pines

J. Gordon Melton, *Encyclopedia of American Religion,* reported by CNN.com, January 31, 2003.

11. Einstein's Brain

Albert Einstein, "The World As I See It," *Living Philosophies* (New York: Simon & Schuster, 1931), pp. 3–7.

Einstein's brain now resides in the Department of Pathology at the Medical Center in Princeton, N.J.

13. What Is a Memory?

Research on cognitive effects of physical abuse reported in *Science News,* June 22, 2002, vol. 161, p. 389.

"Old memories are the result of accumulations of synaptic changes in the cortex as a result of multiple reinstatements of the memory." Joseph LeDoux, *The Synaptic Self* (New York: Touchstone, 1996), p. 107.

Steven Rose, *The Anatomy of Memory,* ed. James McConkey (New York: Oxford University Press, 1996), p. 57.

Carl Gustav Jung, *Modern Man in Search of a Soul,* trans. W. S. Dell and Cary F. Baynes (London: Routledge and Kegan Paul and New York: Harcourt Brace and Company); quoted in *The Creative Process,* ed. Brewster Ghiselin, p. 237.

14. Reflections in a Gazing Ball

Concerning preemptive amnesia, see Steve Madis, "Forget Your Pain," *Brain-Work* (May/June 2003).

Unconscious memory and knowing *how* something happened is sometimes called *implicit* or *procedural* memory.

Jeff Victoroff, *Saving Your Brain* (New York: Bantam, 2002), p. 29.

15. Remember What?

George MacDonald, *Phantastes* (Grand Rapids, Mich.:Wm. B. Eerdmans Publishing Co., 2000), p. 7.

Tip-of-the-tongue memory lapses. See Deborah M. Burke et al., "Learning, Memory and Cognition," *Journal of Experimental Psychology*, November 2002.

When I grasped for the word *awning*, I unknowingly re-created one part of a study in which people were asked to remember certain words that they were tested on later. Those with tip-of-the-tongue memory lapses were shown a list of similar-sounding words, a device that primed their verbal pump, improving their odds of recalling an elusive tip-of-the-tongue word by 25 percent to 50 percent. Hence the success of the "sounds like . . ." strategy when playing charades.

Spencer Nadler, *The Language of Cells* (New York: Random House, 1997).

Fred Cohen, of the University of California at San Francisco, quoted in Sandra Blakeslee, "In Folding Proteins, Clues to Many Diseases," *The New York Times*, May 21, 2002, F1.

Eric R. Kandel and Larry R. Squire, *Memory: From Mind to Molecules* (New York: Scientific American Library, 1999), p. 102.

At Cold Spring Harbor, the neurobiologist Karel Svaboda and his colleagues have been peering into living brains as memories form. They fitted genetically engineered mice (whose neurons make a protein that fluoresces green) with tiny windows over the part of the brain sensitive to what the whiskers detect. The young mice moused for a month, and a camera recorded the flaring greens of their brains. Tiny spines sprouted on the neurons as the mice explored, played, and learned. Some spines retracted within hours; others lingered for months. According to Svaboda, the spines are probably building new synapses, which remain if they're useful, and wither if not. The durable spines may be etching long-term memories. The easy give-and-take is what's surprising, that "neurons are constantly exploring alternative arrangements," building and unbuilding, rewiring themselves daily.

In related experiments at NYU, the neurobiologist Wen-Biao Gan and his colleagues, also studying the birth of spines in mice brains, found that 96 percent of them lingered for at least a month, long enough to engrave lasting memories. "Say a ten-year-old kid uses 1,000 connections to store a piece of information," Gan explains. "When he is eighty, one quarter of the connections will still be there, no matter how things change. That's why you can still remember your childhood experiences."

16. Remember, I Dream

The dream researcher Jonathan Winson, of Rockefeller University in New York City, hypothesizes that "REM sleep may perform a special function in infants. A leading theory proposes that it stimulates nerve growth. Whatever the purpose in infants may be, I suggest that at about the age of two . . . REM sleep takes on its interpretive memory function . . . the concept of the real world against which later experiences must be compared and interpreted." "The Meaning of Dreams," *Scientific American* Special Edition, "The Hidden Mind," spring/summer 2002, p. 60.

Ibid., p. 54.

Neuroscientist Bruce McNaughton attached one hundred electrodes to a rat's hippocampus, monitored the rat's travels, and recorded the brain patterns unique to each location the rat visited. In REM sleep, the patterns repeated in the correct order of the rat's travels. But in non-REM sleep, the pattern was out of order. That suggests that in REM sleep the rat might be generating episodic memories, while other memories form during non-REM sleep. It suggests that the hippocampus keeps replaying the day's patterns during sleep, eventually teaching them to the brain.

17. "Hello," He Lied

Chapter title borrowed from Lynda Obst's memoir about being a film producer in Hollywood.

Brain scan lie detector research by Lawrence Farwell, chief scientist at Brain Fingerprinting Laboratories in Fairfield, Iowa, and Lockheed Martin scientist John Norseen reported in *The New York Times Magazine*, December 9, 2001, p. 82.

18. Traumatic Memories

Jorge Luis Borges, *Labyrinths* (New York: New Directions, 1988), p. 106.

At the American Association for the Advancement of Science annual meeting, held in Denver, Colorado, in February 2003, Matthew Friedman, director of the Vermont-based National Center for Post-Traumatic Stress Disorder, reported on a series of experiments he conducted with Green Berets at Fort Bragg. Soldiers were subjected to harrowing levels of fear. Long, realistic ordeals produced enough battle terror to traumatize other people. But the soldiers' blood work revealed unusually high levels of neuropeptide A, a brain chemical thought to reduce stress and prevent post-traumatic stress disorder in people lucky enough to produce it in abundance. I don't suppose it will be long before the military screens recruits for the protective chemical, allowing only those with enough neuropeptide A to bother training for elite corps.

"Normally we pay attention to the emotionally significant events, including traumatic experiences, that activate the amygdala, so they are also stored as explicit memories in the hippocampus and cerebral cortex. But there is some evidence that extremely high activation of the amygdala shuts down the hippocampus and prevents declarative memories from forming. That could explain why traumatic experiences are explicitly recalled in most cases but sometimes retained only in non-conscious form." "Memories Lost and Found—Part II," *The Harvard Mental Health Letter* 16, no. 2 (August 1999): 2.

Steven A. Mitchell, *Can Love Last?* (New York: W.W. Norton, 2002), p. 199.

"From Instantaneous to Eternal," *Scientific American*, September 2002, pp. 56–57.

19. Smell, Memory, and the Erotic

Marcel Proust, *The Complete Short Stories of Marcel Proust*, trans. Joachim Neugroschel (New York: Cooper Square Press, 2001).

Marguerite Holloway, "The Ascent of Scent," *Scientific American*, December 28, 2001.

20. Introducing the Self

Mitchell, *Can Love Last?*, p. 36.

Virginia Woolf, *Orlando* (New York: Harvest Books, 1993), p. 207.

Difference between *rapture* and *ecstasy*: *Rapture* means, literally, being "seized by force," as if one were a prey animal who is carried away. Caught in the talons of a transcendent rapture, one is gripped, elevated, and trapped at a fearsome height. To the ancient Greeks, this feeling often foretold malevolence and danger—other words that drink from the same rapturous source are *rapacious, rabid, ravenous, ravage, rape, usurp, surreptitious*. Birds of prey that plunge from the skies to gore their victims are known as *raptors*. Seized by a jagged and violent force, the enraptured are carried aloft to their ultimate doom.

Ecstasy also means to be gripped by passion, but from a slightly different perspective: rapture is vertical, ecstasy horizontal. Rapture is high flying, ecstasy occurs on the ground. For some reason, the ancient Greeks were obsessed with the symbol of standing and relied on that one image for countless ideas, feelings, and objects. As a result, a great many of our words today simply reflect where or how things stand: *stanchion, status, stare, staunch, steadfast, statute*, and *constant*. But there are also hundreds of unexpected ones, such as *stank* (standing water), *stallion* (standing in a stall), *star* (standing in the sky), *restaurant* (standing place for the wanderer), *prostate* (standing in front of the bladder), and so on. To the Greeks, *ecstasy* meant to stand outside oneself. How is that possible? Through existential engineering. "Give me a place to stand," Archimedes proclaimed in the third century B.C., "and I will move the earth." Levered by ecstasy, one springs out of one's mind. Thrown free of one's normal self, one stands in another place, at the limits of body, society, and reason, watching the known world dwindle in the *distance* (a spot standing far away).

Legal and ethical dilemmas related to the brain. See "Neuroethics: Mapping the Field," Conference Proceedings, May 13–14, 2002, San Francisco, Calif., (Washington, D.C.: Dana Press, 2003).

21. The Other Self

Shunryu Suzuki, "Posture," in *Zen Mind, Beginner's Mind* (New York: Weatherhill, 1970).

The special immune system cells that collect the invaders are called Langerhans cells.

Gerald N. Callahan, *Faith, Madness, and Spontaneous Combustion* (New York: St. Martin's Press, 2002), pp. 10, 11, 15.

22. Personality

Martha Denckla, "Men, Women, and the Brain," PBS, produced by WETA, Washington, D.C., transcript provided by the Dana Alliance for Brain Initiatives.

Lobotomy surgeon quoted in Edward Shorter, *A History of Psychiatry: From the Era of the Asylum to the Age of Prozac* (New York: John Wiley and Sons, 1977), p. 228.

Chemical rewards of cooperation. MRI study conducted by Clinton Kilts, of Emory University in Atlanta. Reported in *Brain in the News,* July 31, 2002, p. 4.

"unfinished animals" A good discussion of this appears in *The Healing Brain,* by Robert Ornstein and David Sobel (Cambridge, Mass.: Malor Books, 1999), p. 133.

Henry Miller, *Tropic of Capricorn* (New York: Grove Press, 1961), p. 220.

"The child is father of the man," William Wordsworth, "My Heart Leaps Up."

Ellen Ruppel Shell, "Interior Designs," *Discover,* December 2002, p. 51.

Louis Cozolino, *The Neuroscience of Psychotherapy* (New York: W.W. Norton, 2002), pp. 16, 26–27, 49.

Work on serotonin genes by Terrie Moffitt and her team, King's College, London, reported in *Science*, vol. 301, p. 386.

LeDoux quoted in John Fauber, "Work on the Brain Gives Scientists More Insight into Human Body," *Milwaukee Journal Sentinel*, July 8, 2002, p. G1.

Allan N. Shore, clinical psychiatrist and behavioral scientist of the UCLA School of Medicine, in a keynote address delivered to the American Psychological Association, Division of Psychoanalysis, spring meeting, in Los Angeles, 1995.

Richard E. Nisbett, *The Geography of Thought: How Asians and Westerners Think Differently . . . and Why* (New York: Free Press, 2003).

Oxytocin. Josie Glausiusz, "Wired for a Touch," *Discover*, December 2002, p. 13.

23. *"Shall It Be Male or Female? Say the Cells"*

Dylan Thomas, "If I were tickled by the rub of love," in *Collected Poems* (New York: New Directions, 1957), p. 13.

Leonard Shlain, *The Alphabet Versus the Goddess* (New York: Viking, 1998), p. viii.

Roundup of studies reported on in Sharon Lerner, "Good and Bad Marriage, Boon and Bane to Health," *The New York Times*, October 22, 2002, p. F5.

Rotterdam study led by Elisabeth F.C. van Rossum, reported in the October 2002 issue of *Diabetes*.

The Complete Plays of Sophocles, trans. Sir Richard Claverhouse Jebb (New York: Bantam Classic Edition, 1967), p. 131.

SUNY-Stonybrook and Stanford University study, led by Turhan Canli, reported in the July 23, 2002, issue of *Proceedings of the National Academy of Sciences*.

Wiring of female brain. Paul Recer, AP, July 22, 2002. Also reported in "News from the Frontier," *BrainWork*, July-August 2002, p 7.

Women and synesthesia. Richard Cytowic, "Touching Tastes, Seeing Smells—and Shaking Up Brain Science," *Cerebrum* 4, no. 3 (summer 2002), p. 8.

Ruben Gur, professor of psychology, of the University of Pennsylvania, from transcript of "Men, Women and the Brain," PBS, produced by WETA, Washington, D.C., transcript provided by the Dana Alliance for Brain Initiatives.

Maureen Dowd, "Men: Too Emotional?" *New York Times*, July 24, 2002.

MRI rhyming experiments with boys and girls conducted by Drs. Sally and Bennet Shaywitz, Yale University.

Canli et al., *Proceedings of the National Academy of Sciences*, July 23, 2002.

24. Creating Minds

Henri Poincaré, "Mathematical Creation," *The Creative Process* (Berkeley, Calif.: University of California Press, 1984), p. 24.

Semir Zeki, "Artistic Creativity and the Brain," *Science Magazine* online. Zeki is a professor of neurobiology at University College London and cohead of the Wellcome Department of Cognitive Neurology.

Some research suggests that people with schizophrenia have misplaced neurons in the hippocampus or cerebral cortex; other research suggests that schizophrenics have too many excitatory neurons, and poorer filters, and thus are swamped by too many sensations. Geneticist Daniel Cohen and his team at Genset S.A. in Evry, France, have identified two genes that produce proteins they think may interfere with NMDA receptors on brain cells, disrupting the flow of the neurotransmitter glutamate, producing hallucinations and confusion. People with both genes would be at higher risk, and the genes may be specific to primates. Only about 1 percent of children develop schizophrenia if neither parent had it. If one parent did, it becomes 13 percent. If both parents did, that rises to 35 percent. The predisposition for schizophrenia is clearly hereditary, but that doesn't doom. Even though identical twins share the same genes, one twin has only a 30–50 percent chance of developing the disease if the other already has it. Stress seems to play a role in triggering the symptoms. Most schizophrenics (85 percent) smoke heavily, which may be a form of self-medicating. Brain imaging reveals that nicotine animates a part of the brain that's depleted in schizophrenics, and it may help them focus. Schizophrenics have trouble filtering out all the background sensations that we all must in order to function. Smoking seems to help them block some of the background noise. *Brain in the News,* February 2003, p. 2.

25. The Emotional Climate

Linda Gregerson, "Eyes Like Leeks," in *Waterborne* (Boston: Houghton Mifflin, 2002), pp. 2–3.

Patricia Churchland, from "Neuroconscience: Reflections on the Neural Basis of Morality," talk presented at the "Neuroethics: Mapping the Field" conference, held in San Francisco, May 13–14, 2002, p. 48.

Heinrich Heine: "Life is the best teacher, but the tuition is high."

Gerard Manley Hopkins, "No Worst," in *Gerard Manley Hopkins: The Major Works* (New York: Oxford University Press, 2002), p. 167.

26. The Pursuit of Happiness

"Much research indicates that we have a bias for happy memories." For an overview of many studies conducted under various conditions (including floating in a sensory deprivation chamber), see W. Richard Walker, John J. Skowronski, and Charles P. Thompson, "Life Is Pleasant—and Memory Helps to Keep It That Way!" *Review of General Psychology* 7, no. 2 (2003): 203–10.

Antonio Damasio, Richard Davidson, and Jerome Kagan, quoted in "Gray Matters: Emotions and the Brain," National Public Radio, produced in association with the Dana Alliance for Brain Initiatives, March 2000.

Theodore Reik, *Surprise and the Psycho-Analyst* (New York: Dutton, 1937), p. 63.

Robert Browning, "Andrea del Sarto" (1855), l. 97 in *Selected Poems* (New York: Penguin, 2001), p. 101.

Steven Johnson, "Laughter," *Discover,* April 2003, p. 64.

Robert Provine, *Laughter: A Scientific Investigation* (New York: Penguin, 2000).

Sylvia H. Cardoso, "Our Ancient Laughing Brain," *Cerebrum* 2, no. 4 (fall 2000).

B. Knutson, J. Burgdorf, and J. Panksepp, "Anticipation of Play Ellicits High-Frequency Ultrasonic Vocalizations in Young Rats," *Journal of Comparative Psychology* 112 (1998): 65–73.

William M. Kelley, neuroscientist at Dartmouth, and MRI research with volunteers who watched *Seinfeld*.

Brain-imaging data presented at the Society for Neuroscience meeting in Orlando, Florida, during the week of November 10, 2002.

Psychologist Daniel M. Wegner, of Harvard, quoted in John Horgan, "More than Good Intentions: Holding Fast to Faith in Free Will," *The New York Times*, December 31, 2002, p. F3.

27. Memory's Accomplice

Broca's area and bilingual learning studies reported by Judy Foreman, "The Evidence Speaks Well of Bilingualism's Effect for Kids," *Los Angeles Times*, October 7, 2002, p. 1.

"Learning language can begin surprisingly early . . ." See "Gray Matters: The Developing Brain," produced by the Dana Alliance for Brain Initiatives and National Public Radio, 2000.

Tom Costello's e-mail from Slovakia shared with me by his mother, Ann Costello.

Jorge Luis Borges, *Selected Non-Fictions*, ed. Eliot Weinberger (New York: Viking Press, 1999), p. 231.

Russell Hurlburt, "Telling What We Know: Describing Inner Experience," *Trends in Cognitive Neurosciences* 5, no. 9 (September 2001): 4000–3.

28. Metaphors Be with You

Benjamin Lee Whorf, *Language, Thought and Reality*, ed. John B. Carroll (Cambridge, Mass.: MIT Press, 1973), pp. 55–56, 146–48.

George Lakoff and Mark Johnson, *Metaphors We Live By* (Chicago: University of Chicago Press, 1980), pp. 193, 234.

31. Oasis

"the spirit of inhabitable awe": Edward Hirsch, *The Demon and the Angel* (New York: Harcourt, 2002), p. 2.

2.4 billion years ago: Time of Earth acquiring an oxygen-rich atmosphere. See David Catling, "A Breath of Fresh Air," *Seti Institute News,* Second Quarter 2002, vol. 11, no. 2, p. 3. Other estimates put it closer to 3.5 billion years. The Earth is 4.6 billion years old, which means that life arose quickly on the planet, and that its window of opportunity was narrow. As Catling observes, "Oxygen combines the highest energy for metabolism with sufficient stability to be a free atmospheric gas. . . . Had Earth taken a couple of billion years longer to develop this oxygen atmosphere, we wouldn't be here."

Grasshoppers: Conversations with neurobiologist and chemical ecologist Thomas Eisner, of Cornell University, at Highland Hammock and Archbold Research Station, in central Florida's endangered scrublands.

Richard Restak, "The Great Cerebroscope Controversy," *Cerebrum* 2, no. 2 (spring 2000): 24.

32. Conscience and Consciousness

Michael Eigen, *The Psychoanalytic Mystic* (London and New York: Free Association Books, 1998), p. 17.

Patricia Churchland, from "Neuroconscience: Reflections on the Neural Basis of Morality," talk presented at the "Neuroethics: Mapping the Field" conference, held in San Francisco, May 13–14, 2002.

Steven Mithen, *The Prehistory of the Mind* (London: Thames and Hudson, 1996), p. 194.

Stephen M. Kosslyn and Olivier Koenig, *Wet Mind: The New Cognitive Neuroscience* (New York: Free Press, 1992), p. 432.

Gilles Fauconnier and Mark Turner, *The Way We Think* (New York: Basic Books, 2002), pp. 15, 33–34.

33. A Kingdom of Neighbors

Richard Conniff, "Monkey Wrench," *Smithsonian*, pp. 97–104.

Michael Lyvers, *Psyche* 5, no. 31 (December 1999). Dept. of Psychology, Bond University, Gold Coast QLD 4229, Australia.

For more about "the dark parts of the genome," see W. Wayt Gibbs, "The Unseen Genome: Gems Among the Junk," *Scientific American*, November 2003, pp. 47–53.

Michael Pollan, "An Animal's Place," *The New York Times Magazine*, November 10, 2002.

Eugene Linden, *The Octopus and the Orangutan* (New York: Dutton, 2002).

A.A.S. Weir, J. Chappell, A. Kacelnik, "Shaping of Hooks in New Caledonian Crows," *Science*, August 9, 2002.

Graziano, Yap, and Gross, "Coding of Visual Space by Premotor Neurons," *Science*, October 24, 1994, pp. 1054–57.

Larry R. Squire, *Neuron*, April 10, 2003.

We share around 95 percent of our genes with chimpanzees. Roy Britten, of the California Institute of Technology, reported on his research in the *Proceedings of the National Academy of Sciences*. Cited in "Chimps: Not Such Close Relatives," *The Week*, October 11, 2002, p. 24. Using a special computer program, Britten compared 780,000 base pairs of genes in humans and chimps and concluded that there were many more mismatches than the previous estimate of around 99 percent suggested.

Chimpanzee intelligence and emotions: Jane Goodall, "Essays on Science and Society: Learning from Chimpanzees: A Message Humans Can Understand," *Science*, December 18, 1998, pp. 2184–85.

34. The Beautiful Captive

Michael Ondaatje, *The English Patient* (New York: Knopf, 1992), p. 261.

Adrian Raine, "Murderous Minds: Can We See the Mark of Cain?" *Cerebrum* 1, no. 1 (spring 1999): 15–30.

Jonathan Cohen, "Just What's Going on Inside That Head of Yours?" Interview by Sandra Blakeslee, *The New York Times*, March 14, 2000, p. F6.

Brain mapping techniques are fascinating, and the Dana Alliance for Brain Initiatives offers an excellent and readable guide to them online in *Brain Facts: A Primer on the Brain and Nervous System.* www.dana.org.

Some believe obsessive-compulsive disorder may stem from a malfunction of the caudate nucleus, a brain region vital to filtering out mischievous thoughts and impulses.

PET scan studies conducted by Lewis Baxter, of the UCLA School of Medicine, reported by Richard A. Friedman, "Like Drugs, Talk Therapy Can Change Brain Chemistry," *The New York Times*, August 27, 2002, p. F5.

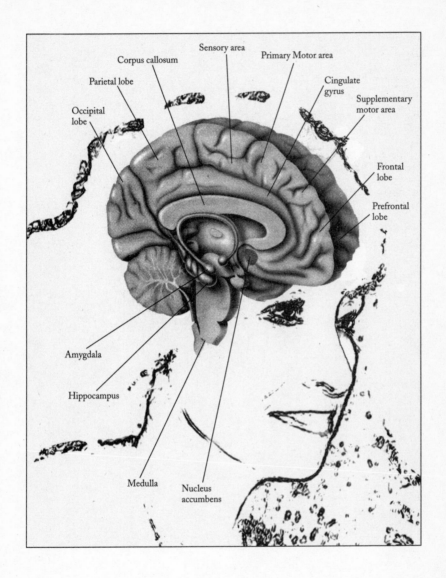

Sensory area

Corpus callosum

Primary Motor area

Parietal lobe

Cingulate gyrus

Occipital lobe

Supplementary motor area

Frontal lobe

Prefrontal lobe

Amygdala

Hippocampus

Medulla

Nucleus accumbens

SELECTED BIBLIOGRAPHY

Ackerman, Diane. *A Natural History of the Senses.* New York: Random House, 1991.
———. *A Slender Thread.* New York: Random House, 1997.
———. *Deep Play.* New York: Random House, 1999.
Alkon, Daniel. *Memory's Voice: Deciphering the Mind-Brain Code.* New York: HarperCollins, 1992.
Alper, Matthew. *The "God" Part of the Brain.* New York: Rogue Press, 1996.
Baron-Cohen, Simon. *The Essential Difference: The Truth about the Male and Female Brain.* New York: Basic Books, 2003.
Bloom, Floyd E., et al. *The Dana Guide to Brain Health.* New York: Free Press, 2003.
Bloom, Floyd E., and Arlyne Lazerson, *Brain, Mind and Behavior,* 2d ed. New York: W. H. Freeman & Co., 1988.
Bollas, Christopher. *The Shadow of the Object: Psychoanalysis of the Unthought Known.* New York: Columbia University Press, 1987.
Borges, Jorge Luis. *Selected Non-Fictions.* Ed. Eliot Weinberger. New York: Viking Press, 1999.
Brockman, John, ed. *The Next Fifty Years: Science in the First Half of the Twenty-first Century.* New York: Vintage, 2002.
Bromberg, Philip M. *Standing in the Spaces: Essays on Clinical Process, Trauma, and Dissociation.* Hillsdale, N.J.: The Analytic Press, 1998.
Browning, Robert. *"Andrea del Sarto" Selected Poems.* New York: Penguin, 2001.
Bruner, Jerome. *Making Stories: Law, Literature, Life.* New York: Farrar, Straus and Giroux, 2002.
Callahan, Gerald N. *Faith, Madness, and Spontaneous Combustion.* New York: St. Martin's Press, 2002.
Carter, Rita. *Exploring Consciousness.* Berkeley: University of California Press, 2002.
———. *Mapping the Mind.* Berkeley: University of California Press, 1999.
Chang, Laura, ed. *The New York Times Scientists at Work.* New York: McGraw-Hill, 2000.
Chase, Stuart. *Power of Words.* New York: Harcourt, Brace, and World, 1953.
Chomsky, Noam. *On Nature and Language.* Cambridge, U.K.: Cambridge University Press, 2002.

Chung, Wilson Chee Jen, "Sexual Differentiation of the Human and Rodent Forebrain." Thesis, University of Amsterdam, 2003.

Cioran, E. M. *A Short History of Decay*. Trans. Richard Howard. New York: Viking, 1975.

Claxton, Guy. *Hare Brain, Tortoise Mind: How Intelligence Increases When You Think Less*. Hopewell, N.J.: Ecco Press, 1997.

Cohen, Jonathan. "Just What's Going on Inside That Head of Yours?" Interview by Sandra Blakeslee, *The New York Times*, March 14, 2000.

Conlan, Roberta, ed. *States of Mind: New Discoveries about How Our Brains Make Us Who We Are*. New York: Dana Press, 1999.

Cozolino, Louis. *The Neuroscience of Psychotherapy*. New York: W. W. Norton, 2002.

Crick, Francis. *The Astonishing Hypothesis: The Scientific Search for the Soul*. New York: Simon & Schuster, 1994.

cummings, e. e., *Complete Poems 1913–1962*. New York: Harcourt Brace Jovanovich, 1963.

Cytowic, Richard E. "Touching Tastes, Seeing Smells—and Shaking Up Brain Science." *Cerebrum* 4, no. 3 (summer 2002): 7–26.

———. *Synesthesia: A Union of the Senses*. 2d ed. Cambridge, Mass.: MIT Press, 2002.

Damasio, Antonio. *Looking for Spinoza: Joy, Sorrow, and the Feeling Brain*. Orlando, Fla.: Harcourt, 2003.

———. *Descartes' Error: Emotion, Reason, and the Human Brain*. New York: Putnam, 1994.

———. *The Feeling of What Happens: Body and Emotion in the Making of Consciousness*. New York: Harcourt Brace, 1999.

DePaulo, J. Raymond, Jr. *Understanding Depression*. Foreword by Kay Redfield Jamison. New York: Dana Press and John Wiley and Sons, 2002.

Eigen, Michael. *The Psychoanalytic Mystic*. London and New York: Free Association Books, 1998.

———. *Damaged Bonds*. London: H. Karnac Books, 2001.

Eisner, Thomas. *For Love of Insects*. Cambridge, Mass.: Harvard University Press, 2003.

Fauconnier, Gilles, and Mark Turner. *The Way We Think: Conceptual Blending and the Mind's Hidden Complexities*. New York: Basic Books, 2002.

Feinberg, Todd E. *Altered Egos: How the Brain Creates the Self*. New York: Oxford University Press, 2001.

Flanagan, Owen. *Consciousness Reconsidered*. Cambridge, Mass.: MIT Press, 1992.

Freud, Sigmund. *The Standard Edition of the Complete Psychological Works of Sigmund Freud*. Ed. and trans. James Strachey. London: Hogarth Press, 1953–74.

Gardner, Howard. *Frames of Mind: The Theory of Multiple Intelligences.* New York: Basic Books, 1983.

Gass, William. *Tests of Time.* New York: Knopf, 2002.

Gawande, Atul. *Complications: A Surgeon's Notes on an Imperfect Science.* New York: Holt, 2002.

Gazzaniga, Michael S. *The Mind's Past.* Berkeley: University of California Press, 1998.

Gazzaniga, Michael, and Todd Heatherton, eds. *Psychological Science: Mind, Brain, and Behavior.* New York: W. W. Norton, 2003.

Ghiselin, Brewster, ed. *The Creative Process.* Berkeley: University of California Press, 1984.

Goleman, Daniel. *Emotional Intelligence.* New York: Bantam Books, 1995.

Gray, J. A., et al. "Possible questions of synesthesia for the hard question of consciousness." In S. Baron-Cohen and J. E. Harrison, eds., *Synaesthesia: Classic and Contemporary Readings.* Oxford: Blackwell, 1997.

Gregerson, Linda. *Waterborne.* Boston: Houghton Mifflin, 2002.

Hamer, Dean, and Peter Copeland. *Living with Our Genes: Why They Matter More Than You Think.* New York: Doubleday, 1998.

Hayakawa, S. I., and Alan R. Hayakawa. *Language in Thought and Action.* San Diego: Harcourt, 1978.

Hirsch, Edward. *The Demon and the Angel: Searching for the Source of Artistic Inspiration.* New York: Harcourt, 2002.

Hobson, J. Allan. *The Chemistry of Conscious States: How the Brain Changes Its Mind.* Boston: Little, Brown, 1994.

Hopkins, Gerard Manley. *Gerard Manley Hopkins: The Major Works.* New York: Oxford University Press, 2002.

Horgan, John. *Rational Mysticism: Dispatches from the Border Between Science and Spirituality.* Boston: Houghton Mifflin, 2003.

Jackendoff, Ray. *Foundations of Language: Brain, Meaning, Grammar, Evolution.* New York: Oxford University Press, 2002.

Jaynes, Julian. *The Origin of Consciousness in the Breakdown of the Bicameral Mind.* Boston: Houghton Mifflin, 1990.

Jones, Judy, and William Wilsom. *An Incomplete Education.* New York: Ballantine Books, 1987.

Kosslyn, Stephen M., and Olivier Koenig. *Wet Mind: The New Cognitive Neuroscience.* New York: Free Press, 1992.

Lakoff, George, and Mark Johnson. *Metaphors We Live By.* Chicago: University of Chicago Press, 1980.

LeDoux, Joseph. *The Emotional Brain.* New York: Touchstone, 1996.

———. *Synaptic Self.* New York: Viking Penguin, 2002.

Linden, Eugene. *The Octopus and the Orangutan.* New York: Dutton, 2002.

Llinás, Rodolfo R. *I of the Vortex: From Neurons to Self.* Cambridge, Mass.: MIT Press, 2002.

MacDonald, George. *Phantastes.* Introduction by C. S. Lewis. Grand Rapids, Mich.: Wm. B. Eerdmans, 2000.

Marcus, Stephen, ed. *Neuroethics: Mapping the Field.* Conference Proceedings. New York: Dana Press, 2002.

Matthews, Paul M., and Jeffery McQuain. *The Bard on the Brain: Understanding the Mind Through the Art of Shakespeare and the Science of Brain Imaging.* New York: Dana Press, 2003.

McConkey, James, ed. *The Anatomy of Memory.* New York: Oxford University Press, 1996.

McEwen, Bruce, with Elizabeth Norton Lasley. *The End of Stress As We Know It.* Foreword by Robert Sapolsky. Washington, D.C.: Dana Press and Joseph Henry Press, 2001.

Miller, Henry. *Tropic of Capricorn.* New York: Grove Press, 1961.

Mitchell, Stephen A. *Can Love Last?* New York: W. W. Norton, 2002.

Mithen, Steven. *The Prehistory of the Mind.* London: Thames and Hudson, 1996.

Nabokov, Vladimir. *Speak, Memory.* New York: Vintage Books, 1989.

Nadler, Spencer. *The Language of Cells: Life as Seen Under the Microscope.* New York: Random House, 2001.

Nisbett, Richard E. *The Geography of Thought: How Asians and Westerners Think Differently . . . and Why.* New York: Free Press, 2003.

Nunn, J. A., et al. "Functional magnetic resonance imaging of synesthesia: activation of V4/V8 by spoken words." *Nature Neuroscience* 5, no. 4 (2002).

Ondaatje, Michael. *The English Patient.* New York: Knopf, 1992.

Ornstein, Robert, and David Sobel. *The Healing Brain.* Cambridge, Mass.: Malor Books, 1999.

Paulesu, E., et al. "The physiology of coloured-hearing: a PET activation study of colour-word synaesthesia." *Brain* 118 (1995).

Pennebaker, James W., ed. *Emotion, Disclosure, and Health.* Washington, D.C.: American Psychological Association, 1995.

Peterson, Brenda, and Toni Frohoff, eds. *Between Species: Celebrating the Dolphin-Human Bond.* San Francisco: Sierra Club Books, 2003.

Pinker, Steven. *The Blank Slate: The Modern Denial of Human Nature.* New York: Viking Penguin, 2002.

————. *The Language Instinct: How the Mind Creates Language.* New York: William Morrow, 1994.

Proust, Marcel. *The Complete Short Stories of Marcel Proust.* Trans. Joachim Neugroschel. New York: Cooper Square Press, 2001.

Provine, Robert. *Laughter: A Scientific Investigation.* New York: Penguin, 2000.

Quartz, Steven R., and Terrence J. Sejnowski. *Liars, Lovers, and Heroes: What the New Brain Science Reveals about How We Become Who We Are.* New York: William Morrow, 2002.

Raine, Adrian. "Murderous Minds: Can We See the Mark of Cain?" *Cerebrum* 1, no. 1 (spring 1999).

Reik, Theodore. *Surprise and the Psycho-Analyst.* New York: Dutton, 1937.

Restak, Richard. "The Great Cerebroscope Controversy." *Cerebrum* 2, no. 2 (spring 2000).

————. *Mysteries of the Mind.* Washington, D.C.: National Geographic, 2000.

————. *The Secret Life of the Brain.* Washington, D.C.: Dana Press and Joseph Henry Press, 2001.

Ridley, Matt. *Nature Via Nurture.* New York: HarperCollins, 2003.

Rilke, Rainer Maria. *Where Silence Reigns: Selected Prose.* New York: New Directions, 1978.

Sapolsky, Robert M. *Why Zebras Don't Get Ulcers: An Updated Guide to Stress, Stress-related Diseases, and Coping.* New York: W. H. Freeman, 1998.

Schachter, Daniel L., and Elaine Scarry, eds. *Memory, Brain, and Belief.* Cambridge, Mass.: Harvard University Press, 2000.

Schachter, Daniel L. *The Seven Sins of Memory: How the Mind Forgets and Remembers.* Boston: Houghton Mifflin, 2001.

Shakespeare, William. *The Complete Works.* London: Arden Shakespeare, 1998.

Shorter, Edward. *A History of Psychiatry: From the Era of the Asylum to the Age of Prozac.* New York: John Wiley and Sons, 1977.

Slaughter, Malcolm M. *Basic Concepts in Neuroscience.* New York: McGraw-Hill, 2002.

Sophocles. *The Complete Plays of Sophocles.* Trans. Sir Richard Claverhouse Jebb. New York: Bantam Classic Edition, 1967.

Squire, Larry R., Floyd E. Bloom, et al., eds. *Fundamental Neuroscience,* 2d ed. San Diego: Academic Press, 2002.

Squire, Larry R., and Eric R. Kandel. *Memory: From Mind to Molecules.* New York: Scientific American Library, 1999.

Stern, Donnel B. *Unformulated Experience: From Dissociation to Imagination in Psychoanalysis.* Hillsdale, N.J.: Analytic Press, 1997.

Storr, Anthony. *The Dynamics of Creation.* New York: Ballantine, 1993.

Strauch, Barbara. *The Primal Teen: What New Discoveries About the Teenage Brain Tell Us About Our Kids.* New York: Doubleday, 2003.

Suzuki, Shunryu. *Zen Mind, Beginner's Mind.* New York: Weatherhill, 1970.

Turkington, Carol. *The Brain Encyclopedia.* New York: Checkmark Books, 1996.

Vertosick, Frank T., Jr. *The Genius Within: Discovering the Intelligence of Every Living Thing.* New York: Harcourt, 2002.

Victoroff, Jeff. *Saving Your Brain: The Revolutionary Plan to Boost Brain Power, Improve Memory, and Protect Yourself Against Aging and Alzheimer's.* New York: Bantam Books, 2002.

Vincent, Jean-Didier. *The Biology of Emotions.* Trans. John Hughes. Oxford: Basil Blackwell, 1990.

Wegner, Daniel M. *The Illusion of Conscious Will.* Cambridge, Mass.: MIT Press, 2002.

Whorf, Benjamin Lee. *Language, Thought and Reality.* Ed. John B. Carroll. Cambridge, Mass.: MIT Press, 1973.

Wilson, Edward O. *The Future of Life.* New York: Knopf, 2002.

Wilson, Timothy D. *Strangers to Ourselves: Discovering the Adaptive Unconscious.* Cambridge, Mass.: Belknap Press, Harvard University Press, 2002.

Wise, Stephen M. *Drawing the Line: Science and the Case for Animal Rights.* Cambridge, Mass.: Perseus Books, 2002.

Woolf, Virginia. *Orlando.* New York: Harvest Books, 1993.

ACKNOWLEDGMENTS

In a thriving garden, no deed is accomplished alone. The same is true of this book whose growth spanned years and required hard gardening, but always nourished my curiosity and fed my sense of wonder.

I'm grateful to the John Simon Guggenheim Foundation for their timely support.

Special thanks to neuroscientists Michael S. Gazzaniga, Larry R. Squire, Barbara Finlay, and psychologist Harry Segal, who were kind enough to read the manuscript for infelicities. The Dana Alliance for Brain Initiatives, especially Jane Nevins of Dana Press, were generous with time and spirit. I also found their publications and online resources helpful and inspiring.

I couldn't have written this book without the aid, comfort, and encouragement of friends and loved ones. Heartfelt thanks especially to Walter Anderson, Philip Bromberg, Ann and John Costello, Tom Costello, Whitney Chadwick, Margaret Dieter, Persis Drell, Thomas Eisner, Rebecca Godin, Lee Kravitz, Jeanne Mackin, Brenda Peterson, Steven Poleskie, William Safire, Dava Sobel, and Paul West. I'm grateful for the guidance of my wonderful agent Virginia Barber and editor Sarah McGrath.

Brief essays inspired by chapters of this book first appeared in *Parade, Parnassus: Poetry in Review, O* magazine, and *The Bard on the Brain*. A little of the Introduction appeared framed (arranged with a CATSCAN and original watercolor) in "Mature Content: Women Artists Look at Aging," an exhibition curated by Rebecca Godin, in Ithaca, New York, November 7–28, 2003.

INDEX

ABOUT THE AUTHOR

Diane Ackerman received an M.A., M.F.A., and Ph.D. from Cornell University. Poet, essayist, and naturalist, she is the author of ten books of adult literary nonfiction, six volumes of poetry, and several nonfiction children's books. Her essays about nature and human nature have appeared in *The New York Times*, *The New Yorker*, *Parade*, *National Geographic*, and other publications. To learn more about her and her books, go to www.dianeackerman.com.